Take Another Look

RETEACHING WORKBOOK

Harcourt Brace & Company

rlando • Atlanta • Austin • Boston • San Francisco • Chicago • Dallas • New York • Toronto • London

http://www.hbschool.com

CONTENTS

Modeling Addition or Subtraction Situations

Using a model helps you know whether to add or subtract.

	Model	**Number Sentence**

Tina had 14 pairs of scissors. She handed out 8 pairs to class-mates. How many pairs of scissors does she have now?

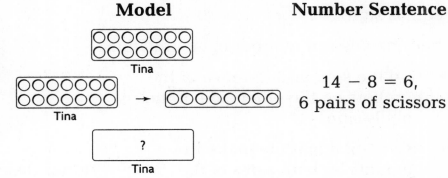

$14 - 8 = 6$,
6 pairs of scissors

Tina collected 6 pairs of scissors. Then she collected 8 more pairs. How many pairs of scissors has she collected?

$6 + 8 = 14$,
14 pairs of scissors

Write the letter of the model that matches the problem.
Then write a number sentence for each problem.

1. Lee had 5 paper clips. Sara gave him 6 more. Now how many paper clips does Lee have?

2. Lee has 13 paper clips. Sara has 6 paper clips. How many more paper clips does Lee have than Sara?

3. Lee had 11 paper clips. He gave 6 to Sara. How many paper clips does he have now?

___ ___   ___ ___   ___ ___

Using Mental Math

If two amounts are equal, they are the same.

An **equation** is a number sentence that uses an equals sign.

To find the missing number in an equation,

$3 + 7 = 2 + \underline{\ ?\ }$

- find the sum or difference of the numbers on the other side of the equals sign.

$10 = 2 + \underline{\ ?\ }$

- use mental math to make the amounts on both sides of the equals sign the same.

Think: Does $10 = 2 + 6$? No
Does $10 = 2 + 8$? Yes

So, $3 + 7 = 2 + 8$.

Complete the steps.

1. $4 + 3 = \underline{\ ?\ } + 6$

$\boxed{} = \underline{\ ?\ } + 6$

$\boxed{} = \boxed{} + 6$

So, $4 + 3 = \boxed{} + 6$.

2. $12 - \underline{\ ?\ } = 9 - 4$

$12 - \underline{\ ?\ } = \boxed{}$

$12 - \boxed{} = \boxed{}$

So, $12 - \boxed{} = 9 - 4$.

3. $0 + 8 = 11 - \underline{\ ?\ }$

$\boxed{} = 11 - \underline{\ ?\ }$

$\boxed{} = 11 - \boxed{}$

So, $0 + 8 = 11 - \boxed{}$.

4. $\underline{\ ?\ } + 6 = 14 - 5$

$\underline{\ ?\ } + 6 = \boxed{}$

$\boxed{} + 6 = \boxed{}$

So, $\boxed{} + 6 = 14 - 5$.

Use mental math to complete each equation.

5. $13 - 6 = \boxed{} + 5$

6. $\boxed{} + 1 = 4 + 4$

7. $12 - 8 = \boxed{} - 9$

Adding Three or More Addends

The **Grouping Property** states that if you group addends differently, the sum does not change.

What is the perimeter of this figure?

Add to find the perimeter. $24 + 18 + 24 + 36 = $ _?_ ft

- Add two numbers at a time.

$$\begin{array}{r} 24 \\ +18 \\ \hline 42 \end{array} \qquad \begin{array}{r} 24 \\ +36 \\ \hline 60 \end{array}$$

- Add the sums.

$$\begin{array}{r} 42 \\ +60 \\ \hline 102 \text{ ft} \end{array}$$

So, the perimeter of this figure is 102 ft.

Find the perimeter of each figure. Then write a number sentence to show how you solved the problem.

1.

$16 + 13 + 16 + 13 = $ _?_ in.

2.

$18 + 24 + 36 = $ _?_ m

3.

$22 + 19 + 11 + 19 = $ _?_ cm

4.

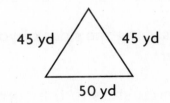

$45 + 45 + 50 = $ _?_ yd

Problem-Solving Strategy

Make a Model

You will need paper clips and counters. Read the problem and then follow the directions. One paper clip equals one yard.

Mr. Taft's sixth-grade class is setting up the gym for a dance. The perimeter of the gym is 80 yards. It is in the shape of a rectangle. The gym is 24 yards long. How wide is it?

Fill in what you know about the dance floor.

1. perimeter _____

2. number of sides _____

3. length of each of 2 sides _____

Take 80 paper clips. Build the 2 sides whose lengths you know.

4. How many paper clips are left? _____

Take the remaining paper clips. Build 2 sides of equal length.

5. How wide is the gym?

_____

The decoration committee is going to arrange clusters of balloons along the perimeter of the dance floor. There will be 4 clusters of balloons on each side, including 1 at each corner.

Place counters along the sides of your model to represent balloon clusters.

6. Count the clusters in the corners. How many? _____

7. Count the clusters not in the corners. How many? _____

8. How many balloon clusters will be placed in all?

_____

Estimating Sums and Differences

Round 63 to the nearest ten.

You can use a number line to help you.

- Mark 63 on the number line.
- Circle the two nearest tens.
- Decide which ten 63 is closer to. Round to that number.
- Since 63 is closer to 60, you round to 60.

Remember: When a number is halfway between two numbers, round to the greater number.

Round to the nearest ten. Use the number line.

1.

20   25   30

$28 \rightarrow$ _____

2.

80   85   90

$84 \rightarrow$ _____

3.

90   95   100

$91 \rightarrow$ _____

4.

40   45   50

$47 \rightarrow$ _____

Round to the nearest hundred. Use the number line.

5.

100   150   200

$159 \rightarrow$ _____

6.

500   550   600

$510 \rightarrow$ _____

Estimate the sum or difference. Use the rounded numbers from Exercises 1–6.

7. $\begin{array}{r} 28 \rightarrow \\ +47 \rightarrow \\ \hline ? \end{array}$ ____

8. $\begin{array}{r} 91 \rightarrow \\ -84 \rightarrow \\ \hline ? \end{array}$ ____

9. $\begin{array}{r} 47 \rightarrow \\ +91 \rightarrow \\ \hline ? \end{array}$ ____

10. $\begin{array}{r} 84 \rightarrow \\ -28 \rightarrow \\ \hline ? \end{array}$ ____

11. $\begin{array}{r} 159 \rightarrow \\ +510 \rightarrow \\ \hline ? \end{array}$ ____

12. $\begin{array}{r} 510 \rightarrow \\ -159 \rightarrow \\ \hline ? \end{array}$ ____

Adding and Subtracting with Money

When you write money amounts,

- leave two places for cents after the decimal point.
- place a dollar sign in front of the money amount.

925
↓
9.25
$| 9.25

Write the money amount.

A. 805 → $| 8 . 0 5

B. 10 → $| 0 . 1 0

1. 346 → ☐| __ . __ __

4. 65 → ☐| __ . __ __

2. 100 → ☐| __ . __ __

5. 43 → ☐| __ . __ __

3. 598 → __ __ . __ __

6. 17 → __ __ . __ __

C. 1482 → $| 1 4 . 8 2

D. 6 → $| 0 . 0 6

7. 2337 → ☐| __ __ . __ __

10. 3 → ☐| __ . __ __

8. 1164 → ☐| __ __ . __ __

11. 5 → ☐| __ . __ __

9. 2815 → __ __ . __ __

12. 9 → __ __ . __ __

Find the sum or difference.

13. $5.41
   + 3.72

14. $17.66
   + 11.42

15. $9.83
   − 7.91

16. $24.15
   − 18.60

17. $12.87
   + 7.09

18. $20.50
   − 5.29

19. $23.75
   − 10.99

20. $2.29
    3.05
   + 1.81

Look at Exercises 13–20. Circle *yes* or *no.*

21. There are two places for cents after each decimal point.   yes   no

22. There is a dollar sign in front of each money amount.   yes   no

Adding and Subtracting Larger Numbers

What is the length of the fourth side of this figure?

You can use addition *and* subtraction to find the missing length of a figure.

- Add the known sides.
- Subtract the sum from the perimeter.

$$
\begin{array}{r}
19 \\
38 \\
+\ 38 \\
\hline
\boxed{95}
\end{array}
\qquad
\begin{array}{r}
143 \\
-\ \boxed{95} \\
\hline
48
\end{array}
$$

? = 48 ft

38 ft   38 ft

19 ft

The length of the fourth side is 48 feet.

Perimeter = 143 ft

Use addition *and* subtraction to find the missing length.

1.   124 yd

102 yd   102 yd

?

Perimeter = 404 yd

$$
\begin{array}{r}
\underline{} \\
+\ \underline{} \\
\hline
\boxed{}
\end{array}
\qquad
\begin{array}{r}
404 \\
-\ \boxed{} \\
\hline
\end{array}
$$

? = _____

2.   ?

29 ft   29 ft

Perimeter = 87 ft

$$
\begin{array}{r}
\underline{} \\
+\ \underline{} \\
\hline
\boxed{}
\end{array}
\qquad
\begin{array}{r}
87 \\
-\ \boxed{} \\
\hline
\end{array}
$$

? = _____

3.   24 m
48 m
?
48 m
24 m

Perimeter = 202 m

$$
\begin{array}{r}
\underline{} \\
\underline{} \\
+\ \underline{} \\
\hline
\boxed{}
\end{array}
\qquad
\begin{array}{r}
202 \\
-\ \boxed{} \\
\hline
\end{array}
$$

? = _____

4.   ?

32 ft   16 ft

Perimeter = 88 ft

? = _____

5.   88 m

33 m   ?

88 m

Perimeter = 242 m

? = _____

6.   37 yd   21 yd

?

37 yd   21 yd

Perimeter = 140 yd

? = _____

Subtracting Across Zeros

To subtract across zeros, it is helpful to make a model with base-ten blocks.

Can you take 127 from 5 hundreds? No. Regroup 1 hundred as 10 tens.

$$\begin{array}{r} 500 \\ -127 \\ \hline \end{array}$$

Can you take 127 from 4 hundreds, 10 tens? No. Regroup 1 ten as 10 ones.

$$\begin{array}{r} \overset{4}{\cancel{5}} \ \overset{10}{\cancel{0}} \ 0 \\ -1 \ \ 2 \ \ 7 \\ \hline \end{array}$$

Can you take 127 from 4 hundreds, 9 tens, and 10 ones? Yes.

$$\begin{array}{r} \overset{9}{} \\ \overset{4}{\cancel{5}} \ \overset{\cancel{10}}{\cancel{0}} \ \overset{10}{\cancel{0}} \\ -1 \ \ 2 \ \ 7 \\ \hline 3 \ \ 7 \ \ 3 \end{array}$$

Hundreds	Tens	Ones
5 hundreds		
4 hundreds	10 tens	
4 hundreds	9 tens	10 ones

Complete Exercises 1–4. Regroup 1 hundred as 10 tens and 1 ten as 10 ones.

1. 3 hundreds = ____ hundreds ____ tens ____ ones

2. 6 hundreds = ____ hundreds ____ tens ____ ones

3. 4 hundreds = ____ hundreds ____ tens ____ ones

4. 9 hundreds = ____ hundreds ____ tens ____ ones

Regroup. Find the difference.

5. $\begin{array}{r} 300 \\ -158 \\ \hline \end{array}$

6. $\begin{array}{r} 600 \\ -149 \\ \hline \end{array}$

7. $\begin{array}{r} 400 \\ -325 \\ \hline \end{array}$

8. $\begin{array}{r} 900 \\ -276 \\ \hline \end{array}$

More Subtracting Across Zeros

When you add or subtract numbers with many digits, it's important that you write the numbers carefully so that digits with the same place value line up. This also helps you write regrouped digits in the correct places.

$$
\begin{array}{r}
\overset{3}{\cancel{4}}\overset{9}{\cancel{0}},\overset{9}{\cancel{0}}\overset{9}{\cancel{0}}\overset{10}{\cancel{0}} \\
-\ 3\,2,9\,6\,3 \\
\hline
7,0\,3\,7
\end{array}
$$

Copy each exercise into the space below. Find the difference.
Remember to leave the first row of boxes empty for regrouping.

1. 900
    − 831

2. 5,002
    − 2,764

3. 7,001
    − 3,208

4. 6,000
    − 1,955

5. 80,000
    − 31,523

6. 10,000
    − 4,661

7. 40,000
    − 20,118

8. 70,000
    − 55,049

9. 60,000
    − 33,604

Problem-Solving Strategy

Work Backward

Sometimes you can work backward to solve a problem.

Bruce collected some sand dollars. He gave 10 to his younger sister. His friend, Susan, gave him 7 sand dollars. He put 4 back in the water. He had 14 sand dollars left. How many did he collect?

Operation	Opposite Operation
Add	Subtract
Subtract	Add
Divide	Multiply
Multiply	Divide

- First, write the steps of the problem in order from left to right. Use numbers and operations to describe the steps.

Number of sand dollars collected.	He gave 10 to sister.	Susan gave him 7.	He put 4 back.	Number left
__?__ →	__− 10__ →	__+ 7__ →	__− 4__ →	__14__

- Then, reverse the order and use the opposite operations. Find the answer.

$$\underline{21} \leftarrow \underline{+\,10} \leftarrow \underline{-\,7} \leftarrow \underline{+\,4} \leftarrow \underline{14}$$

So, Bruce collected 21 sand dollars.

Reverse the order and use opposite operations. Find the answer.

1. $\underline{\quad?\quad} \rightarrow \underline{+\,3} \rightarrow \underline{-\,8} \rightarrow \underline{-\,11} \rightarrow \underline{27}$

$\underline{\quad\quad} \leftarrow \underline{\quad\quad} \leftarrow \underline{\quad\quad} \leftarrow \underline{\quad\quad} \leftarrow \underline{\quad\quad}$

2. $\underline{\quad?\quad} \rightarrow \underline{-\,11} \rightarrow \underline{+\,25} \rightarrow \underline{52}$

$\underline{\quad\quad} \leftarrow \underline{\quad\quad} \leftarrow \underline{\quad\quad} \leftarrow \underline{\quad\quad}$

3. Sasha and Chuck played a number game. Sasha picked a number and subtracted 3. Next, she added 12. Then, she subtracted 20. Last, she added 11. The result was 30. What number did Sasha begin with?

_____

Estimating Sums and Differences

An estimate is an answer that is close to the exact answer.

Round 4,782 to the nearest *thousand*.

Follow these rules:

- Circle the digit to be rounded.   ④782

- Underline the digit to its right.   ④782

- If the underlined digit is 5 or more, the circled   7 > 5, 5,xxx
  digit increases by 1. If the underlined digit is less
  than 5, the circled digit stays the same.
                                                              ↓
- Write zeros in all places to the right of the   5,000
  circled digit.

So, 4,782 rounded to the nearest thousand is 5,000.

Circle the digit to be rounded. Underline the digit to its right.
Round 9,385 to the given place value.

1. nearest ten   **2.** nearest hundred   **3.** nearest thousand

9,385   9,385   9,385

_____   _____   _____

Round 2,617 to the given place value.

4. nearest ten   **5.** nearest hundred   **6.** nearest thousand

2,617   2,617   2,617

_____   _____   _____

Round each number to the greatest place value. Write the
estimated sum or difference.

7.   8 9 1 → ☐☐☐   **8.**   3,456 → ☐,☐☐☐
       − 6 2 3 → − ☐☐☐   + 7,099 → + ☐,☐☐☐
       ─────────────────   ──────────────────
           ? →   ☐☐☐   ? → ☐☐,☐☐☐

Name _____

LESSON 2.5

Choosing the Operation

In deciding if you should add or subtract to solve a problem, think about the answer. Should the answer be greater than the numbers in the problem? If yes, then consider adding. If no, then consider subtracting.

Eldrick bought a book for $6.95. He gave the clerk $10.00. What should his change be?

Think: Should the answer be more than or less than $10.00?

_____

Solve: $10.00
            −   6.95
           —————————
             $3.05

Read the problem. Answer the **Think** question. Then solve.

1. Devon bought a card for $1.75 and a magazine for $2.95. How much did he spend?

Think: Should the answer be more than or less than $2.95?

_____

Solve:

2. Alicia paid $8.25 for a calendar with horse pictures on it. Becky paid $11.50 for a calendar with cat pictures on it. How much more did Becky pay than Alicia?

Think: Should the answer be more than or less than $11.50?

_____

Solve:

3. Scott had $15.00 when he went to the store. He has $4.39 left. How much money did he spend at the store?

Think: Should the answer be more than or less than $15.00?

_____

Solve:

R12   TAKE ANOTHER LOOK

What Questions Can Multiplication Answer?

When groups *are not* equal, you add to find how many. When groups *are* equal, you can add or multiply to find how many.

Rita bought 3 packages of batteries. There were 2 batteries in each package. How many batteries did Rita buy?

Draw a picture:

Equal groups? __yes__

Add: __2 + 2 + 2 = 6__

Multiply: 3 × 2 = 6

Rita bought 3 packages of batteries. Two packages had 2 batteries each. The third package had 3 batteries. How many batteries did Rita buy?

Draw a picture:

Equal groups? __no__

Add: __2 + 2 + 3 = 7__

Multiply: cannot

Decide if the groups are equal. Add to find the sum. Multiply if you can. If you cannot multiply, write *cannot*.

1.

Equal groups? _____

Add: _____

Multiply: _____

2.

Equal groups? _____

Add: _____

Multiply: _____

3.

Equal groups? _____

Add: _____

Multiply: _____

Draw a picture. Add to find the sum. Multiply if you can. If you cannot multiply, write *cannot*.

4. Alberto handed out crayons. He gave Pete 5 crayons, Max 7 crayons, and Leslie 8 crayons. How many crayons did Alberto hand out? Draw a picture:

Add: _____

Multiply: _____

5. Jyoti handed out crayons. She gave 5 crayons to each student in the first row. There are 4 students in the first row. How many crayons did Jyoti hand out? Draw a picture:

Add: _____

Multiply: _____

Multiplication Properties

In a multiplication sentence, the numbers being multiplied are the factors; the answer is the product.

$2 \times 5 = 10$
factors product

$5 \times 2 = 10$
factors product

$9 \times 0 = 0$
factors product

Sort. Write each multiplication sentence in the column under the property it shows.

$8 \times 0 = 0$	$1 \times 5 = 5$	$4 \times 1 = 4$	$3 \times 6 = 18$
$9 \times 0 = 0$	$0 \times 4 = 0$	$0 \times 6 = 0$	$2 \times 0 = 0$
$1 \times 8 = 8$	$3 \times 0 = 0$	$3 \times 1 = 3$	$5 \times 0 = 0$
$6 \times 1 = 6$	$4 \times 5 = 20$	$0 \times 7 = 0$	$9 \times 1 = 9$
$7 \times 2 = 14$	$1 \times 1 = 1$	$1 \times 7 = 7$	

1.
Property of One

When one of the factors is 1, the product is the other factor.

2.
Zero Property

When one of the factors is zero, the product is zero.

3.
Order Property

Two numbers can be multiplied in any order. The product is the same.

_____

$1 \times 2 = 2$

_____

_____

_____

_____

_____

_____

$0 \times 1 = 0$

_____

_____

_____

_____

_____

$3 \times 4 = 12$

$4 \times 3 = 12$

$2 \times 7 = 14$

_____

$6 \times 3 = 18$

_____

$5 \times 4 = 20$

_____

More About Multiplication Models

You can use the facts for 5 and a model to help you find
the product of factors greater than 5. What is 7 × 8?

5 Facts Chart

$7 \times 8 = \underline{\ ?\ }$

Step 1   Write an addition expression
breaking apart one factor.

$5 \quad + \quad 2$

$(5 \times 8) \quad + \quad (2 \times 8) = \underline{\ ?\ }$

Step 2   Use the 5 Facts Chart and a
model to find the products.

Step 3   Find the sum of the products.

$40 \quad + \quad 16 = 56$

So, $7 \times 8 = 56$.

Complete each step. Use the 5 Facts Chart to help you draw
a model.

1. $9 \times 6 = \underline{\ ?\ }$

$5 \quad + \quad 4$

$(5 \times 6) + (4 \times 6) =$

$\underline{\quad\quad} + \underline{\quad\quad}$

So, $9 \times 6 = \underline{\quad\quad}$

2. $8 \times 9 = \underline{\ ?\ }$

$5 \quad + \quad 3$

$(5 \times 9) + (3 \times 9) =$

$\underline{\quad\quad} + \underline{\quad\quad}$

So, $8 \times 9 = \underline{\quad\quad}$

3. $6 \times 7 = \underline{\ ?\ }$

$(5 \quad + \quad 2)$

$(6 \times 5) + (6 \times 2) =$

$\underline{\quad\quad} + \underline{\quad\quad}$

So, $6 \times 7 = \underline{\quad\quad}$

4. $8 \times 6 = \underline{\ ?\ }$

$(5 \quad + \quad 3)$

$(5 \times 6) + (3 \times 6) =$

$\underline{\quad\quad} + \underline{\quad\quad}$

So, $8 \times 6 = \underline{\quad\quad}$

Multiplying Three Factors

The **Grouping Property** may help you when you multiply. Remember when the grouping of factors is changed, the product remains the same. Look at the example below which shows the Grouping Property.

$(3 \times 2) \times 2$ gives the same answer as $3 \times (2 \times 2)$.

Choose the way of grouping that is easier for you to multiply. What is $5 \times 2 \times 3$?

- Group the factors two ways. $(5 \times 2) \times 3 = \underline{\ ?\ }$ or $5 \times (2 \times 3) = \underline{\ ?\ }$

- Choose the way that is easier for you.

- Solve.   **Think:** $5 \times 2 = 10$. It's easy to multiply tens. $10 \times 3 = 30$ or
$5 \times 6 = 30$

Circle the grouping that is easier for you. Show how you solve it.

1. $(2 \times 2) \times 3 = \underline{\ ?\ }$,
or $2 \times (2 \times 3) = \underline{\ ?\ }$

$\underline{\hspace{3cm} 2 \times 6 = 12 \hspace{3cm}}$

2. $(2 \times 5) \times 3 = \underline{\ ?\ }$,
or $2 \times (5 \times 3) = \underline{\ ?\ }$

$\underline{\hspace{5cm}}$

3. $(4 \times 2) \times 3 = \underline{\ ?\ }$,
or $4 \times (2 \times 3) = \underline{\ ?\ }$

$\underline{\hspace{5cm}}$

4. $(5 \times 3) \times 3 = \underline{\ ?\ }$,
or $5 \times (3 \times 3) = \underline{\ ?\ }$

$\underline{\hspace{5cm}}$

5. $(2 \times 2) \times 8 = \underline{\ ?\ }$,
or $2 \times (2 \times 8) = \underline{\ ?\ }$

$\underline{\hspace{5cm}}$

6. $(4 \times 3) \times 3 = \underline{\ ?\ }$,
or $4 \times (3 \times 3) = \underline{\ ?\ }$

$\underline{\hspace{5cm}}$

7. $(5 \times 2) \times 2 = \underline{\ ?\ }$,
or $5 \times (2 \times 2) = \underline{\ ?\ }$

$\underline{\hspace{5cm}}$

8. $(2 \times 6) \times 1 = \underline{\ ?\ }$,
or $2 \times (6 \times 1) = \underline{\ ?\ }$

$\underline{\hspace{5cm}}$

Problem-Solving Strategy

Make a Model

In Mr. Paul's classroom there are 3 rows of tables.
There are 2 tables in each row. Each table has 4 chairs.
How many chairs are in Mr. Paul's classroom?

You can *make a model* to find the number of chairs in
the classroom.

Use tiles to make a model.

Row 1

Row 2

Row 3

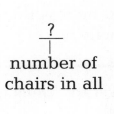

$$(3 \quad \times \quad 2) \quad \times \quad 4 \quad = \quad ?$$

number of   number of   number of   number of
  rows   tables   chairs at   chairs in all
                           each table

$6 \times 4 = 24$

So, there are 24 chairs in Mr. Paul's classroom.

Complete each step. Make a model. Find the product.

1. Ms. Carter uses a bulletin board
to display students' work. The
bulletin board has 4 rows of
papers on it. There are 5 papers
in each row. How many papers
are there in all?

    number of rows _____

    number of papers in each

    row _____

    number of bulletin boards _____

    (_____ × _____) × _____ = _?_

    _____

    So, there are _____ papers in all.

2. In P.E. class the students are
standing in 6 rows. There are 4
students in each row. The P.E.
teacher gives 2 hand weights to
each student. How many hand
weights does she give out?

    number of rows _____

    number of students in each

    row _____

    number of hand weights for

    each student _____

    _____ × (_____ × _____) = _?_

    _____

    So, she gives out _____ weights.

Choosing the Operation

In deciding if you need to add, multiply, or subtract to solve a problem, think about the answer.

• If the answer should be greater than the numbers in the problem, **add.**

• If the answer should be greater than the numbers in the problem, and the problem includes equal groups, **multiply.**

• If the answer should be less than the greater number in the problem, **subtract.**

Read the problem. Answer the **Think** questions. Then solve by using a number sentence.

1. Rosemary bicycled 6 miles before lunch and 8 miles after lunch. How many miles did she bicycle in all?

   Think: Should the answer be greater than or less than

   8 miles?_____

   If greater than, does the problem include equal

   groups? _____

   Solve: _____

   So, Rosemary bicycled _____.

2. Eric had 8 marbles. He gave some to Wendy. Now he has 6 marbles left. How many marbles did he give to Wendy?

   Think: Should the answer be greater than or less than

   8 marbles? _____

   If greater than, does the problem include equal

   groups? _____

   Solve: _____

   So, Eric gave Wendy _____.

What Questions Can Division Answer?

You can use division to find how many are in each group or how many equal groups there are. You can use counters to model the division.

Example

Bill is handing out 20 paint-brushes. He places 4 on each table. On how many tables can he place paintbrushes?

total number _____20_

number in each group _4_

number of groups _____?_

Use counters to find the **number of groups.**

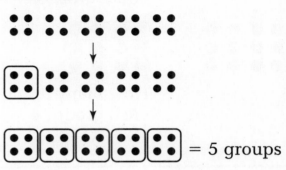

= 5 groups

So, he can place paintbrushes on 5 tables.

Write the information from the problem. Write a question mark to show what you need to find. Then use counters to solve.

1. Eva has 24 beads to make 3 bracelets. Each bracelet will have the same number of beads. How many beads will she use for each bracelet?

   total number _____

   number in each group _____

   number of groups _____

   She will use _____.

2. Pam is gluing photos onto pieces of poster board. She has 27 photos. She is gluing 9 onto each piece of poster board. How many pieces of poster board will she need?

   total number _____

   number in each group _____

   number of groups _____

   She will need _____.

Connecting Multiplication and Division

Multiplication and division are inverse operations. One operation undoes the other. You can use models to show this.

$3 \times 4 = 12$

$12 \div 3 = ?$
Divide 12 into 3 equal groups.

How many are in each group? 4
$12 \div 3 = 4$

$5 \times 2 = 10$

$10 \div 5 = ?$
Divide 10 into 5 equal groups.

How many are in each group? 2
$10 \div 5 = 2$

Circle equal groups to show the division. Complete the division sentence.

1. $6 \times 3 = 18$

$18 \div 6 =$ _____

2. $2 \times 7 = 14$

$14 \div 2 =$ _____

3. $5 \times 4 = 20$

$20 \div 5 =$ _____

4. $3 \times 9 = 27$

$27 \div 3 =$ _____

Complete each multiplication and division sentence. You may wish to use a model to help you.

5. $8 \times 6 =$ _____

$48 \div 8 =$ _____

6. $7 \times 4 =$ _____

$28 \div 7 =$ _____

7. $2 \times 9 =$ _____

$18 \div 2 =$ _____

8. $3 \times 3 =$ _____

$9 \div 3 =$ _____

9. $1 \times 8 =$ _____

$8 \div 1 =$ _____

10. $5 \times 7 =$ _____

$35 \div 5 =$ _____

Dividing with Remainders

Facts of 2	Facts of 3	Facts of 4	Facts of 5	Facts of 6
$18 \div 2 = 9$	$27 \div 3 = 9$	$36 \div 4 = 9$	$45 \div 5 = 9$	$54 \div 6 = 9$
$16 \div 2 = 8$	$24 \div 3 = 8$	$32 \div 4 = 8$	$40 \div 5 = 8$	$48 \div 6 = 8$
$14 \div 2 = 7$	$21 \div 3 = 7$	$28 \div 4 = 7$	$35 \div 5 = 7$	$42 \div 6 = 7$
$12 \div 2 = 6$	$18 \div 3 = 6$	$24 \div 4 = 6$	$30 \div 5 = 6$	$36 \div 6 = 6$
$10 \div 2 = 5$	$15 \div 3 = 5$	$20 \div 4 = 5$	$25 \div 5 = 5$	$30 \div 6 = 5$
$8 \div 2 = 4$	$12 \div 3 = 4$	$16 \div 4 = 4$	$20 \div 5 = 4$	$24 \div 6 = 4$
$6 \div 2 = 3$	$9 \div 3 = 3$	$12 \div 4 = 3$	$15 \div 5 = 3$	$18 \div 6 = 3$
$4 \div 2 = 2$	$6 \div 3 = 2$	$8 \div 4 = 2$	$10 \div 5 = 2$	$12 \div 6 = 2$
$2 \div 2 = 1$	$3 \div 3 = 1$	$4 \div 4 = 1$	$5 \div 5 = 1$	$6 \div 6 = 1$

To find the quotient and remainder, first find the fact whose dividend is closest to yet less than the dividend in the problem. Then subtract the dividends.

What is $23 \div 5$?

fact: $20 \div 5 = 4$ subtract: $23 - 20 = 3$ So, $23 \div 5 = 4$ r3.

Use the table and subtract to find the quotient and remainder.

1. What is $27 \div 6$?

fact: $24 \div 6 = 4$

subtract: _____

So, $27 \div 6 =$ _____.

2. What is $14 \div 4$?

fact: $12 \div 4 = 3$

subtract: _____

So, $14 \div 4 =$ _____.

3. What is $17 \div 3$?

_____

_____

$17 \div 3 =$ _____

4. What is $34 \div 5$?

_____

_____

$34 \div 5 =$ _____

5. What is $15 \div 2$?

_____

_____

$15 \div 2 =$ _____

Problem-Solving Strategy

Make a Table

You can make a table to show all the possible ways something can be done. You can use the information in the table to solve problems.

Charise has decorated 36 barrettes to sell at a craft fair. She needs to decide whether to put 4, 5, 6, or 8 barrettes in each package. She made a table to help find out how she might package the barrettes.

Complete the table.

	Total Number of Barrettes	Number in Each Package	Division Sentence	Number of Packages	Number of Barrettes Left Over
	36	4	$36 \div 4 = 9$	9	0
1.	36	5	_____	___	___
2.	36	6	_____	___	___
3.	36	8	_____	___	___

4. If Charise does not want to have any barrettes left over, how many ways can she package the barrettes? What are the ways?

_____

5. If Charise would like to have 1 barrette left over to keep for herself, how many barrettes should she put in each package?

_____

6. How many more barrettes would Charise need to make to put 8 barrettes in each package and have none left over?

_____

7. Charise decides she wants to make the fewest packages she can and have no leftover barrettes. How many barrettes should she put in each package?

_____

Division on a Multiplication Table

You can use a multiplication table and two index cards to find a quotient. To find the quotient $12 \div 4 = \underline{?}$, follow these steps.

- Write a related multiplication fact with a missing factor.

  $? \times 4 = 12$

- Find the given factor in the top row of the chart. Place an index card alongside the column it is in.

- Find the product 12 in that column. Place an index card below. The first number in that row is the answer.

So, $12 \div 4 = 3$.

Write the related multiplication sentence. Use the multiplication table to solve.

1. $27 \div 3 = \underline{\ ?\ }$

   _____

   $27 \div 3 = \underline{\ \ \ }$

2. $42 \div 7 = \underline{\ ?\ }$

   _____

   $42 \div 7 = \underline{\ \ \ }$

3. $24 \div 6 = \underline{\ ?\ }$

   _____

   $24 \div 6 = \underline{\ \ \ }$

×	0	1	2	3	4	5	6	7	8	9
0	0	0	0	0	0	0	0	0	0	0
1	0	1	2	3	4	5	6	7	8	9
2	0	2	4	6	8	10	12	14	16	18
3	0	3	6	9	12	15	18	21	24	27
4	0	4	8	12	16	20	24	28	32	36
5	0	5	10	15	20	25	30	35	40	45
6	0	6	12	18	24	30	36	42	48	54
7	0	7	14	21	28	35	42	49	56	63
8	0	8	16	24	32	40	48	56	64	72
9	0	9	18	27	36	45	54	63	72	81

Use the multiplication table to find the quotient.

4. $36 \div 4 = \underline{\ \ \ }$

5. $48 \div 8 = \underline{\ \ \ }$

6. $20 \div 5 = \underline{\ \ \ }$

7. $28 \div 7 = \underline{\ \ \ }$

8. $56 \div 7 = \underline{\ \ \ }$

9. $45 \div 9 = \underline{\ \ \ }$

10. $36 \div 6 = \underline{\ \ \ }$

11. $63 \div 9 = \underline{\ \ \ }$

12. $81 \div 9 = \underline{\ \ \ }$

Recording and Practicing Division

You can record division problems in two ways: $6 \div 1 = ?$ or $1\overline{)6}$.

How many counters?	6	••••••
How many groups?	1	
How many in each group?	6	

So, $6 \div 1 = 6$ or $1\overline{)6}$.

Use counters to show the division. Complete each table.

	Division Problem	Number of Counters	Number of Groups	Number in Each Group	Division Sentence
1.	$4 \div 4 = \underline{\ ?\ }$	4	4	1	$4 \div 4 = \underline{\quad}$
2.	$6 \div 6 = \underline{\ ?\ }$	___	___	___	$6 \div 6 = \underline{\quad}$
3.	$7 \div 7 = \underline{\ ?\ }$	___	___	___	
4.	$8 \div 8 = \underline{\ ?\ }$	___	___	___	

5. When a number is divided by itself, what is the quotient? ___

	Division Problem	Number of Counters	Number of Groups	Number in Each Group	Division Sentence
6.	$1\overline{)3}$	3	1		$1\overline{)3}$
7.	$1\overline{)5}$				$\overline{)\ }$
8.	$1\overline{)8}$				$\overline{)\ }$
9.	$1\overline{)9}$				$\overline{)\ }$

10. When a number is divided by 1, what is the quotient? _____

Choosing the Operation

Decide if you should add, multiply, subtract, or divide to get
the answer.

- If the answer should be greater than the numbers
  in the problem, **add.**

- If the answer should be greater than the numbers
  in the problem, and the problem includes equal
  groups, **multiply.**

- If the answer should be less than the greater number
  in the problem, **subtract.**

- If the answer should be less than the greater number
  in the problem, and the problem includes equal
  groups, **divide.**

Read the problem. Answer the **Think** questions. Choose the
operation, and then solve.

1. Lewis bought 6
packs of astronomy
cards. Each pack
had 7 cards in it.
How many cards
did Lewis buy?

Think: Should the answer be greater than or

less than 7 cards? _____

Does the problem include equal groups?

_____

Solve: _____

Lewis bought _____.

2. Hugh, Carol, and
Lee collected 24
pinecones. They
divided them so
that they each got
the same number.
How many
pinecones did
they each get?

Think: Should the answer be greater than or

less than 24 pinecones? _____

Does the problem include equal groups?

_____

Solve: _____

They each got _____.

How Numbers Are Used

Different kinds of numbers give us different information.

Cardinal numbers tell how many.	**Ordinal numbers** tell position or order.	**Nominal numbers** name things.
• **1** inch • **two** hours • **6** students • **10** acorns • **fifty-two** miles	• **4th** grade • **ninth** birthday • **first** anniversary • **18th** floor	• radio station: **103.1** FM • telephone number: **555-4027** • identification number: **007**

In the diagram below, color each cardinal number blue, each ordinal number yellow, and each nominal number green.

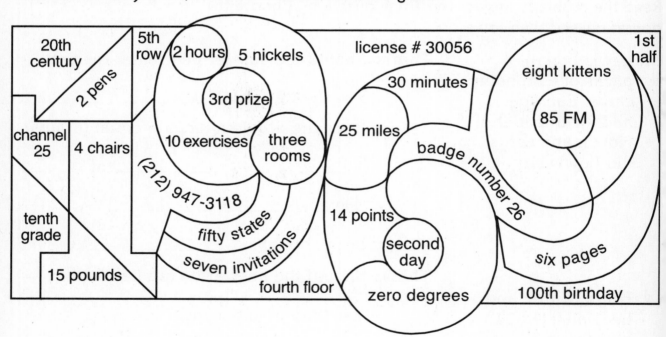

In what year did the first astronaut walk on the moon? Look

at your picture for the answer. _____

Name _____

More About How Numbers Are Used

The two numbers used to locate points on a grid are called
an **ordered pair.**

What is the ordered pair for the
location of the Ferris wheel?

(_?_ , _?_)

⊛ = Ferris wheel

- Locate the Ferris wheel. Follow
  the gridline **down** to a number.
  That is the first number in the
  ordered pair.

- Follow the gridline **across** to a
  number. That is the second
  number in the ordered pair.

    (3, _?_)

    (3, 2)

So, (3, 2) is the ordered pair for the location of the Ferris wheel.

Complete the ordered pair for the location of the Ferris wheel.

1.

(1, ___)

2.

(___, 2)

3.

(___, ___)

Write the ordered pair for each ride.

4. merry-go-round   _____

5. bumper cars   _____

6. roller coaster   _____

7. tea cups   _____

Legend
R = roller coaster
B = bumper cars
M = merry-go-round
T = tea cups

Understanding 1,000

You can represent numbers from 1 to 1,000 using 10-by-10 grids.

Remember: = 1   = 10   = 100

How many 10-by-10 grids do you need to show 438?

Show 438.   **Think:** 438 = 400 + 30 + 8

- Shade grids to show 400.

- Shade columns to show 30.

- Shade squares to show 8.

So, you need 5 grids to show 438.

Write the number. Tell how many 10-by-10 grids are needed.
(Grids that are overlapped are completely shaded.)

1. number _____   2. number _____

   number of grids _____   number of grids _____

Write the number of 10-by-10 grids you would need to show the number.

3. 80   4. 800   5. 808   6. 961

_____   _____   _____   _____

Name _____

More About 1,000

In order to read and write numbers to thousands, it is important to know the value of each digit in a number.

You can write the numbers 6,507 and 2,089 in words.

• Write each digit in its place in a place-value chart.	Thousands	Hundreds	Tens	Ones	Thousands	Hundreds	Tens	Ones
	6	5	0	7	2	0	8	9
• Write the number in expanded form.	6,000 + 500 + 7				2,000 + 80 + 9			
• Write the number in words. Remember to write a comma after *thousand*.	six thousand, five hundred seven				two thousand, eighty-nine			

Write each number in expanded form. Then write each number in words.

1. 8,021 = _____ + _____ + _____

   _____

2. 4,617 = _____ + _____ + _____ + _____

   _____

3. 1,938 = _____ + _____ + _____ + _____

   _____

4. 6,472 = _____ + _____ + _____ + _____

   _____

5. 9,706 = _____ + _____ + _____

   _____

Problem-Solving Strategy

Act It Out

Laverne, Jody, Pearl, Leroy, and Cal are each performing in a variety show. Laverne will perform after Leroy. Pearl will perform third. Jody will perform before Pearl. Cal will perform before Jody. In what order will they perform?

Act it out to solve. Write each person's name on a slip of paper. When you know a person's position, place the slip of paper with his or her name in that position.

Pearl will perform third.

| ? | ? | Pearl | ? | ? |

Jody will perform before Pearl. Cal will perform before Jody.

| Cal | Jody | Pearl | ? | ? |

Laverne will perform after Leroy.

| Cal | Jody | Pearl | Leroy | Laverne |

So, the order is: Cal, Jody, Pearl, Leroy, Laverne.

Act it out to solve.

1. Vicky has 5 classes before lunch each day. She has math before science. She has Spanish before math. She has English right before lunch. She has history after science. In what order does Vicky have classes in the morning?

_____

2. There are 6 children in the Ingraham family. Denise is older than Steven, but younger than Eliot. Kayla is younger than all her brothers and sisters except for Kirk. Bernie is younger than Steven. Write the names of the Ingraham children in order from oldest to youngest.

_____

Benchmark Numbers

You can use a benchmark number to help you estimate the
number of flowers in gardens *A* and *B*.

- Determine how much
space 5 flowers fill.
This is your bench-
mark number.

- Estimate how many times
larger is the space filled
by flowers in each garden.

A   B

The space is about   The space is about
4 times larger.   8 times larger.
$4 \times 5 = 20$ $8 \times 5 = 40$

So, there are about 20 flowers in garden *A* and about
40 flowers in garden *B*.

Complete each problem to estimate the number of seeds shown.

1.

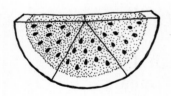

benchmark number _____

_____ $\times$ _____ = _____

about _____ seeds.

2.

benchmark number _____

_____ $\times$ _____ = _____

about _____ seeds.

Estimate the number of buds on the branches.

3.

_____

4.

_____

Numeration Systems

A numeration system is a way of naming numbers and counting numbers. The people of ancient Egypt used these symbols to name numbers.

Egyptian	⚘ Lotus Flower	☌ Scroll	∩ Heel Bone	| Stick
Standard	1,000	100	10	1

The Egyptian numeration system did not use place value, so the position and order of the symbols did not matter.

For example, the number ∩∩∩\|\| and the number ∩∩\|\|∩ both equal the standard number 32.

Write the value of each group of Egyptian symbols.

1. ∩∩\|\|\| _____

2. \|\|\|∩∩ _____

3. ☌☌☌☌\|\| _____

4. ⚘☌∩∩\|\|\|\| _____

Use a table to help you understand the value of each digit in a large number, such as 3,457. The number 3,457 means 3 thousands + 4 hundreds + 5 tens + 7 ones.

Thousands		Hundreds	Tens	Ones
3	,	4	5	7

Complete.

5. In the number 346, the value of the 3 is _____ hundreds.

6. In the number 3,459, the value of the 9 is _____ ones.

Write the value of the digit 5 in each of the following numbers.

7. 502   8. 365   9. 5,403   10. 3,450

_____   _____   _____   _____

Reading and Writing Numbers

You can show numbers using base-ten blocks, or write numbers in different ways.

1 thousand   3 hundreds   2 tens   5 ones

Expanded form $1,000 + 300 + 20 + 5$

Standard form $1,325$

Written form   one thousand, three hundred twenty-five

Use the model to help you write the number in different ways.

1.

Expanded form _____ + _____ + _____

Standard form _____

Written form _____

2.

Expanded form _____ + _____ + _____

Standard form _____

Written form _____

Write each number in expanded form.

3. 541 _____   4. 5,333 _____

Name _____

Mental Math and Place Value

You can use place value to find different names for the
same number.

Find two other names for 140.

140 = ____ hundred + 4 tens   or   140 = ____ tens

Write each number two ways.

1. 160 = ____ hundred ____ tens   2. 180 = ____ hundred ____ tens

   160 = ____ tens   180 = _____ tens

3. 290 = 2 hundreds ____ tens   4. 400 = ____ hundreds

   290 = ____ tens   400 = ____ tens

Use another name for each number. Solve Exercises 5–6 by using
mental math.

5. 80 + 90 = ___?___   6. 700 + 600 = ___?___

   80 → [____] tens   700 → [____] hundreds
   +90 → +[____] tens   +600 → +[____] hundreds
        [____] tens = 170   [____] hundreds = 1,300

Use another name for each number. Solve by using mental math.

7. 30   8. 400   9. 600   10. 9,000   11. 5,000
   +60   +800   +600   +2,000   +8,000

Name _____

Place Value to 100,000

In our place-value system, each place is ten times greater than the place to its right.

Hundred Thousands	Ten Thousands	Thousands	Hundreds	Tens	Ones
100,000	10,000	1,000	100	10	1
10 ten thousands	10 thousands	10 hundreds	10 tens	10 ones	1 one

- What is the next largest place after hundreds?
  Think: It is ten times greater than hundreds, so it must be thousands.

- What is the next largest place after thousands?
  Think: It is ten times greater than thousands, so it must be ten thousands.

Complete the place-value chart for the number 111,111.

Hundred Thousands	_____	Thousands	_____	Tens	Ones
1. _____	10,000	_____	100	_____	1

Complete the following equations.

2. _____ ten thousands = 100,000

3. _____ hundreds = 1,000

4. _____ thousands = 10,000

5. 10 _____ = 10

6. 10 _____ = 100

7. 10 _____ = 1,000

8. ten thousands = _____

9. 100 thousands = _____

TAKE ANOTHER LOOK   R35

Using Large Numbers

Large numbers are separated into periods by commas. A period is a group of three numbers in a large number. The commas make the number easier to read.

Millions				Thousands				Ones		
Hundreds	Tens	Ones		Hundreds	Tens	Ones		Hundreds	Tens	Ones
	4	3	,	3	0	4	,	2	3	5

Read: 43 million, 304 thousand, 235

Write: 43,304,235

Complete to show how to read each number.

1. 45,234,500 _____ million, _____ thousand, _____

2. 8,345,688 _____ million, _____ thousand, _____

3. 240,555,678 240 _____, 555 _____, _____

4. 34,111,800 34 _____, 111 _____, _____

Write each number in standard form.

5. 43 thousand, 345 _____

6. 24 million, 234 thousand, 100 _____

7. 5 million, 8 thousand, 34 _____

Write the digit that is in the ten thousands place.

8. 456,100 _____   9. 1,290,000 _____   10. 123,450,356 _____

Write the digit that is in the hundred thousands place.

11. 256,100 _____   12. 9,876,000 _____   13. 300,123,666 _____

Write the digit that is in the millions place.

14. 5,000,123 _____   15. 346,123,000 _____   16. 301,000,000 _____

Use a Table

A table is helpful in organizing data. Find 1940 under the
Year column on the table. Look across the table to find the
number of farm workers for that year.

Persons in Farm Occupations in the United States, 1940–1990	
Year	Number of Farm Workers
1940	8,995,000
1950	6,858,000
1960	4,132,000
1970	2,881,000
1980	2,818,000
1990	2,864,000

For Problems 1–4, use the table.

1. Were there more farm workers in the United States in

   1940 or in 1990? _____ How many more? _____

2. Between which two decades did the number of farm workers

   increase? _____ and _____

   What was the amount of this increase? _____

3. Do you think that there were more or fewer farm workers

   in 1930 than in 1940? _____

   Why? _____

4. After reading the table, Carlos predicted that the number
   of farm workers in the year 2000 would be about 2,800,000.

   Is this a reasonable prediction? _____

   Why or why not? _____

   _____

Comparing on a Number Line

Locate the numbers 20 and 50 on the number line by drawing a dot for each number.

You can compare 20 and 50 by writing 20 < 50 or 50 > 20.

Locate each pair of numbers on the number line by drawing dots.
Then show two ways to compare the numbers.

1. 70, 90

_____ _____

2. 600, 300

_____ _____

3. 4,000, 2,000

_____ _____

4. 25, 50

_____ _____

5. 34, 37

_____ _____

6. 240, 220

_____ _____

7. 3,500, 3,200

_____ _____

8. 750, 500

_____ _____

Comparing Numbers

Which number is greater, 3,491 or 3,487?

Step 1   Write one number under the other.

Step 2   Compare the digits, beginning with the greatest place-value position.

Step 3   Circle the first digits that are different.

same digits
same digits
↓   ↓
3 , 4 8 7
3 , 4 9 1

Think: Since 8 tens < 9 tens,
3,487 < 3,491.

Circle the greater number in each pair of numbers. Then write < or >.

1. 627
637

627 ◯ 637

2. 129
127

129 ◯ 127

3. 599
600

599 ◯ 600

4. 1,345
1,444

1,345 ◯ 1,444

5. 2,400
3,200

2,400 ◯ 3,200

6. 5,225
4,988

5,225 ◯ 4,988

7. 3,010
3,009

3,010 ◯ 3,009

8. 1,456
998

1,456 ◯ 998

9. 890
894

890 ◯ 894

10. 7,203
7,320

7,203 ◯ 7,320

11. 4,408
4,423

4,408 ◯ 4,423

12. 763
739

763 ◯ 739

13. 410
503

410 ◯ 503

14. 1,062
1,028

1,062 ◯ 1,028

15. 5,301
5,401

5,301 ◯ 5,401

Problem-Solving Strategy

Guess and Check

Sometimes you can find an answer by guessing first and
then checking your answer. Then guess, check, and guess
again until you get the answer.

There are 16 lions and tigers at the zoo. There are 4 more lions
than tigers. How many lions and how many tigers are there?

	Tigers	Lions	Total	Think
Guess 1	4	8	12	not enough animals
Guess 2	7	11	18	too many animals
Guess 3	6	10	16	just right

Use the table to help you guess and check. You may need
to guess more than 3 times.

1. There are 24 plants in Tamara's garden. There are twice
as many tomato plants as pepper plants. How many
tomato and pepper plants are there?

_____

	Pepper Plants	Tomato Plants	Total
Guess 1			
Guess 2			
Guess 3			

2. Leon and Jim have 45 model racing cars. Leon has 5 more
cars than Jim. How many cars does each boy have?

_____

	Jim's Cars	Leon's Cars	Total
Guess 1			
Guess 2			
Guess 3			

Name _____

Ordering Numbers

Cut pieces of adding-machine paper the same length as
each of the following:

- Your <u>foot</u>

- Your <u>hand</u> (from wrist to longest fingertip)

- Your <u>arm</u> (from shoulder to longest fingertip)

- Distance around your <u>ankle</u>

Label each piece of paper with the name of the body part.

Compare and order the pieces of paper from the least to the
greatest. List the measurements in order from the least to the
greatest.

1. _____   2. _____

3. _____   4. _____

The Jones family records the height of each family member on
New Year's Day. One year, they made this table.

Name	Mrs. Jones	Bill	Lisa	Mr. Jones	Tom	Jeff
Height (in inches)	66 in.	54 in.	39 in.	71 in.	35 in.	48 in.

5. List the heights in order from the greatest to the least.

_____

6. How much taller was Bill than Tom? _____

7. How much more did Tom need to grow to be 3 feet tall? _____

8. What was the difference in height between the tallest

and the shortest family members? _____

9. One year later, Lisa was 2 inches taller and Jeff was 1 inch taller.

What was the difference in their heights then? _____

Name _____

More About Ordering Numbers

The table shows the heights of three mountains in Vermont.

Mountain	Height in Feet
Ellen	4,135 ft
Camel's Hump	4,083 ft
Abraham	4,052 ft

List the heights of the mountains in order from the tallest to the shortest.

Step 1
Compare the digits in each place-value position, beginning with the thousands.

4,135
↓
4,083
↓
4,052

Step 2
Compare the digits in the next place-value position.

4,135
↓
4,083
↓
4,052

Think: 1 hundred > 0 hundreds, so 4,135 is the greatest number.

Step 3
Compare 4,083 and 4,052.

4,083
↓
4,052

Think: 8 tens > 5 tens, so 4,083 > 4,052.

So, the heights of the mountains in order from the tallest to the shortest are 4,135; 4,083; 4,052.

Circle the greatest number and underline the least number.

1. 738
689
691

2. 3,670
3,760
3,761

3. 9,456
8,390
8,389

4. 7,803
7,549
7,201

5. 4,023
4,068
4,005

6. 6,260
6,275
6,270

Sorting and Comparing

The librarian recorded the number of books that each student
read during the summer reading program at the library.

Alana	Lily	Mark	Sean	Marco	Molly	Jean	Tom
22 books	7 books	38 books	12 books	9 books	18 books	31 books	6 books

The Venn diagram below shows one way to compare the numbers
of books read.

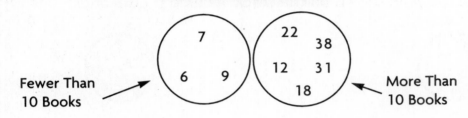

Number of Books Read

Fewer Than
10 Books

More Than
10 Books

1. How many students read fewer than 10 books? _____

2. How many students read more than 10 books? _____

For Exercises 3–4, use the table above. Insert the numbers where
they belong in the Venn diagrams.

3. **Number of Books Read**

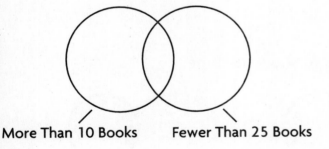

More Than 10 Books   Fewer Than 25 Books

4. **Number of Books Read**

Fewer Than
20 Books

Fewer Than
40 Books

Solve.

5. How many students read more than 10 books but fewer

   than 25 books? _____

6. How many students read fewer than 20 books?

   _____

Name _____

Telling Time

Analog Clock

The *second* hand shows the seconds and moves around the clock in 1 minute.

The *minute* hand is the long hand.

The *hour* hand is the short hand.

Read: 15 minutes and 10 seconds past 5, or 5:15 and 10 seconds

Digital Clock

Read: 35 minutes past 2

Complete to write the time shown on the clock. Include the seconds.

1.

_____ minutes and _____ seconds

past _____

2.

_____ minutes and _____ seconds

past _____

3.

_____ minutes and _____ seconds

past _____

4.

_____ minutes and _____ seconds

past _____

Draw the hour hand and the minute hand to show the time.

5.

3:25

6.

8:05

7.

9:57

Write *second, minute, hour,* or *day.*

8. It took Joe 45 _____ to wash and vacuum his car.

9. A snap of your fingers takes about 1 _____.

10. Anne's piano lesson was 1 _____ long.

A.M. and P.M.

A.M. means "the time between midnight and noon."

P.M. means "the time between noon and midnight."

At 8:15 A.M. Kathy goes to school.

At 8:15 P.M. Kathy goes to bed.

Write A.M. or P.M.

1. George eats supper at 5:30 _____.

2. Carlos goes to school from 8:40 _____ to 2:40 _____.

3. Maria has a violin lesson from 3:45 _____ to 4:30 _____.

Write the time, using A.M. or P.M.

4. the time the library closes

_____

5. the time Kitty sees the sunset

_____

6. the time Lu takes a spelling test at school

_____

7. the time Frank gets home from school

_____

8. the time Jack has soccer practice

_____

9. the time Dad cooks breakfast

_____

Elapsed Time on a Clock

Skip counting can help you tell the elapsed time in minutes.
How many minutes is it

from 10:40 to 11:15?

from 6:15 to 7:00?

It is 35 minutes from 10:40 to 11:15.

It is 45 minutes from 6:15 to 7:00.

Tell how many minutes will elapse.

1. from 12:30 to 12:45

_____

2. from 4:25 to 5:20

_____

3. from 10:15 to 11:05

_____

4. from 11:15 to 11:40

_____

5. from 7:05 to 8:00

_____

6. from 6:35 to 6:45

_____

Using a Schedule

WORKSHOP SCHEDULE			
Class	Days	Start Time	End Time
Computer	Tue or Thu	3:30 P.M.	4:15 P.M.
Cooking	Wed	3:00 P.M.	4:00 P.M.
Leather Crafts	Mon or Thu	3:00 P.M.	4:15 P.M.
Drama	Mon	4:00 P.M.	4:45 P.M.
Pottery	Wed	3:15 P.M.	4:15 P.M.

Rosie wants to take a class after school. She is free on
Mondays. She needs to take a bus home at 4:30 P.M.
What class can she take?

Step 1 Find the column marked *Days*. The leather crafts
and drama classes meet on Mondays.

Step 2 Find the column marked *End Time*. Rosie can take the
leather crafts class because it ends before 4:30 P.M.

For Problems 1–7, use the workshop schedule.

1. Which classes last 1 hour? _____

2. Which class is the longest? _____

3. Which classes could you take on Thursday?

   _____

4. Could you take both cooking and pottery? _____

5. John is taking cooking and computer. How long does

   he attend classes? _____

6. It takes Bill 20 minutes to walk home after pottery class.

   When does he arrive at home? _____

7. One day Jill's father picked her up after leather crafts
   class. He arrived at 4:55 P.M. How long did Jill have to

   wait for her father? _____

Name _____

Problem-Solving Strategy

Make a Table

Saturday Movie Times at the Library

Time Travel will be shown at 9:30 A.M., 11:00 A.M., and 2:15 P.M. Movie length is 45 minutes.

Spaceship Adventure will be shown at 10:30 A.M., 1:15 P.M., and 4:00 P.M. Movie length is 2 hours.

Chen Wong would like to see both movies. He needs to be home by 2:30 P.M.

1. Complete the table, and then circle the times Chen Wong can see each movie. Remember to include A.M. and P.M.

Name of Movie	Starting Times	Ending Times
Time Travel	9:30 A.M. _____	10:15 A.M. _____
_____	_____	_____

For Problems 2 and 3, use the table.

2. Bill walked home from the library after watching the 10:30 A.M. showing of *Spaceship Adventure*. It took him 30 minutes to walk home. When did Bill arrive home?

_____

3. Maria was at the library from 2:00 P.M. to 4:00 P.M. During that time, she watched *Time Travel* and worked on a report. How much time did she spend

working on her report? _____

Elapsed Time on a Calendar

You can use a calendar to find elapsed time. How many days will the Harvest Festival last?

- Look at the October calendar. Find 9.
- Count each day of the festival.

So, the festival will last 10 days.

Wilson Farm began advertising the Harvest Festival 5 weeks before the first day of the festival. On what date did Wilson Farm begin advertising?

- Begin at October 9. It is a Thursday.
- Use the calendars to count back 5 weeks, or 5 Thursdays.

So, Wilson Farm began advertising September 4.

For Problems 1–4, use the calendars.

1. Write the date of the first Sunday during the festival. _____

2. Tom bought a pumpkin on the last day of the festival. How many days will it be from then until October 31?

   _____

3. Julian went to the festival on October 13 after school. What day of the week was that? _____

4. Wilson Farm began harvesting pumpkins 2 weeks before the beginning of the festival. On what date did the farm begin harvesting pumpkins? _____

| September |||||||
Sun	Mon	Tue	Wed	Thu	Fri	Sat
	1	2	3	4	5	6
7	8	9	10	11	12	13
14	15	16	17	18	19	20
21	22	23	24	25	26	27
28	29	30				

| October |||||||
Sun	Mon	Tue	Wed	Thu	Fri	Sat
			1	2	3	4
5	6	7	8	9	10	11
12	13	14	15	16	17	18
19	20	21	22	23	24	25
26	27	28	29	30	31	

Organizing Data in Tables

Mr. Cohen used tally marks to keep a record of the number of gerbils that he sold each week in his pet shop.

Week	Gerbils Sold
1	~~HHt~~ ~~HHt~~ /
2	~~HHt~~ ///
3	~~HHt~~ ///
4	////

1 = 🐀

~~HHt~~ = 🐀🐀🐀🐀🐀

In the frequency table, the *frequency* is the number of gerbils Mr. Cohen sold in a week. The *cumulative frequency* is the number of gerbils sold so far.

Use the tally table to complete the frequency table below.

NUMBER OF GERBILS SOLD		
Week	Gerbils Sold	Cumulative Frequency
1			← sold in Week 1
2			← sold in Weeks 1 and 2
3			← sold in Weeks 1, 2, and 3
4			← sold in Weeks 1, 2, 3, and 4

For Exercises 1–3, use the frequency table to answer the questions.

1. During which week did Mr. Cohen sell the most gerbils?

   _____

2. During which two weeks did Mr. Cohen sell the same

   number of gerbils? _____

3. How many gerbils did Mr. Cohen sell in Weeks 3 and 4?

   _____

Name _____

Problem-Solving Strategy

Make an Organized List

Kyle wants to make a beanbag game to help his younger brother learn about shapes and colors. He will use the colors red, yellow, and blue. He will use circles, squares, and triangles. Kyle made an organized list and drew a picture of each beanbag he can make.

	Red	Yellow	Blue
Circle	R	Y	B
Square	R	Y	B
Triangle	R	Y	B

1. How did Kyle organize his list of possible beanbags?

   _____

2. How many different beanbags can Kyle make using

   3 colors and 3 shapes? _____

3. a. Make an organized list by drawing beanbags for a game that uses these colors and shapes: red, blue, yellow, and green; square and circle.

   b. How many different beanbags can you make?

   _____

Understanding Surveys

In a **survey,** you ask several people the same question and record their answers. Mr. Baker made a survey of favorite kinds of ice cream for the class party.

This table shows the results of Mr. Baker's first survey.

For Exercises 1–4, use the First Survey table.

FIRST SURVEY What is your favorite ice cream flavor?	
Flavor	**Votes**
Chocolate	ℋℋ /
Peppermint	ℋℋ
Strawberry	ℋℋ
Peach	/
Mint Chip	/
Chocolate Chip	///
Vanilla	ℋℋ

1. How many students chose vanilla?

_____

2. How many flavors of ice cream did

the students choose? _____

3. Which flavor got the most votes?

_____

4. How many students chose either mint chip or chocolate chip as their

favorite flavor? _____

Mr. Baker wanted to buy just two flavors of ice cream. He made a second survey to decide which two flavors to buy.

SECOND SURVEY Which of these ice cream flavors do you like best— chocolate, strawberry, peppermint, or vanilla?	
Flavor	**Votes**
Chocolate	ℋℋ ///
Strawberry	ℋℋ /
Peppermint	ℋℋ //
Vanilla	ℋℋ

5. Why do you think Mr. Baker picked these flavors for his second survey?

_____

_____

6. Did more students choose chocolate in the first survey or in the second survey?

_____

7. Explain why you think some flavors got more votes in the second survey.

_____

Comparing Graphs

This vertical bar graph shows how many items were sold one Saturday at the Toy Corner.

How many puzzles were sold?

Follow these steps to help you read the graph.

Step 1 Read across the bottom to find *Puzzles.* Put your finger on the word *Puzzles.*

Step 2 Follow the bar to the top. Place a piece of paper at the top of the bar. Look to the left scale and read.

Step 3 The bar stops halfway between 8 and 10. The number is 9.

There were 9 puzzles sold.

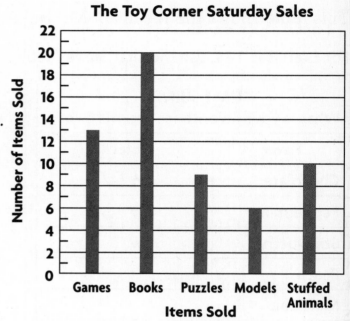

The Toy Corner Saturday Sales

For Exercises 1–10, use the graph.
How many of each item were sold?

1. Games _____
2. Books _____
3. Puzzles _____

4. Models _____
5. Stuffed Animals _____

Which item sold more?

6. Stuffed Animals or Puzzles?
7. Games or Models?
8. Games or Stuffed Animals?

_____ _____ _____

9. How many items were sold in all? _____

10. Which two items added together sold almost the same

   number as books? _____

   _____

More About Comparing Graphs

Tom and his friends earned money cutting lawns during the summer.

Lawns Cut This Summer

Name	Number of Lawns
Tom	6
Jack	10
Barb	8
Kim	4
Lia	3

The *Scale* is the series of numbers placed at fixed distances.

To make a bar graph with different intervals, think about how the graph will look when you are finished. Make two graphs. Follow these steps.

Step 1 Choose an interval for the graph. Label each item and the numbers on the scale.

Step 2 Use a pencil and ruler to measure and fill in the tops, sides, and bottoms of the bars.

Step 3 Check to see if you have correctly filled in the graphs. Color the bars with markers.

Use the table above to complete the bar graphs. Draw a bar to show the number of lawns cut by each person. Use a pencil, ruler, and markers.

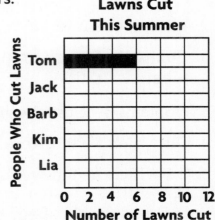

Name _____

Exploring Double-Bar Graphs

Peter made a table of the number of shells that he collected on two beaches. Then he started a **double-bar graph** to compare the data. Each bar represents the number of each kind of shell collected.

SHELLS COLLECTED ON WEST AND DUNE BEACHES

Shell	West Beach	Dune Beach
Clam	6	8
Oyster	9	1
Scallop	10	2
Periwinkle	6	6

Key
■ West Beach
▨ Dune Beach

1. What scale did Peter choose? _____

2. What is the interval of the scale? _____

3. Complete Peter's graph. Remember to include
   • bars.   • title.   • labels.

For Problems 4–5, use the graph.

4. On which beach were more oyster shells collected? _____

5. On which beach were an equal number of clam shells

   and periwinkle shells collected? _____

Name _____

Reading a Line Graph

Mr. King made this line graph to show the number of
bicycles he sold each month for the first six months of the
year. This **line graph** shows how bike sales change over time.

How many bicycles were sold in April?

Step 1 Find the line labeled *April*. Put your finger on
the line. Follow that line up to the point (•).

Step 2 Move your finger along the line to the left to
locate the number of bicycles sold in April.

Mr. King sold 30 bicycles during April.

For Problems 1–5, use the graph.

1. Mr. King sold the greatest number of bicycles during

_____ and _____.

2. During March, Mr. King sold _____ bicycles.

3. During February, Mr. King sold _____ bicycles.

4. Mr. King sold _____ more bicycles in June than in January.

5. There was an increase in the number of bicycles sold from

_____

_____

Line Plot

Ms. Ryan is teaching a cartooning class for students ages 8 to 12. She made a line plot. This **line plot** shows the number of students of each age in the class.

Ages of Students in Class

How many students are 10 years old?

Step 1   Locate 10 on the diagram.

Step 2   Count the number of X's above the number 10.

There are 4 students who are 10 years old.

For Problems 1–4, use the line plot.

1. There are_____ students who are 11 years old.

2. More students are _____ years old than any other age.

3. The same number of students are _____ and _____ years old.

4. The **range** is the difference between the greatest and least numbers in a set of data. The range in ages of

   students in the class is _____.

5. After the first morning of cartooning class, the students counted the number of cartoon pages they had completed. Use the data in the table to make a line plot.

CARTOON PAGES COMPLETED						
Number of Pages	1	2	3	4	5	6
Number of Students	0	2	4	7	2	1

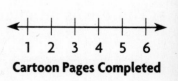

Cartoon Pages Completed

Stem-and-Leaf Plot

Becky measured the height in inches of each sunflower in her garden. She made a list of the heights.

20 inches	14 inches	17 inches
35 inches	12 inches	15 inches
12 inches	32 inches	27 inches
23 inches	24 inches	21 inches
27 inches	27 inches	33 inches
18 inches	33 inches	

- To organize the data in a **stem-and-leaf plot** write the numbers in order from least to greatest. Group the data in rows according to the digit in the tens place.

- To find the heights of the shortest and tallest sunflowers in the garden, look at the first and last numbers in the stem-and-leaf plot.

Stem (Tens)	Leaves (Ones)
1	2 2 4 5 7 8
2	0 1 3 4 7 7 7
3	2 3 3 5

Heights of Sunflowers in Inches

The shortest sunflower is 12 inches. The tallest sunflower is 35 inches.

For Problems 1–5, use the stem-and-leaf plot.

1. How many sunflowers are 33 inches tall? _____

2. How many sunflowers are 20 or more inches tall? _____

3. The **mode** is the number that occurs most often in a set

   of data. What is the mode for the sunflower heights? _____

4. The **median** is the middle number in an ordered series of

   numbers. What is the median sunflower height? _____

5. The range is the difference between the greatest and the least numbers in a data set. What is the range in

   sunflower heights? _____

Name _____

Choosing a Graph

The students in Mr. Schmitt's fourth-grade class participated in a read-a-thon for four weeks. The bar graph shows the number of books of each kind that students read. The line graph shows the number of books students read each week during the read-a-thon.

1. What kind of graph is Graph A? _____

2. What do the bars in Graph A represent? _____

_____

3. Did students read more biographies or more science books? How does the graph show this?

_____

4. What kind of graph is Graph B? _____

5. Did the number of books read each week increase or decrease? How does the graph show this?

_____

6. What would the line connecting the data points look like if the number of books read each week was the same?

_____

7. What other kind of graph could you use to show the

data in Graph A? _____

Problem-Solving Strategy

Make a Graph

During the first week of summer vacation, Andy and Pam opened up a lemonade stand in front of their house. They kept a record of the number of cups of lemonade they sold each afternoon during the week.

Andy and Pam want to show how the sales of lemonade changed during the week.

LEMONADE SALES FOR ONE WEEK	
Day	**Number of Cups Sold**
Monday	8
Tuesday	10
Wednesday	12
Thursday	12
Friday	12
Saturday	20
Sunday	22

Follow these steps to complete the line graph.

Step 1 Use the table to find how many cups of lemonade were sold on Monday.

Step 2 Find on the graph the line labeled Monday.

Step 3 Follow that line up to the intersecting line labeled 8.

Step 4 Place a point on the graph.

Step 5 Repeat Steps 1–4 for the other six days of the week.

Step 6 Use a straightedge to connect the points on the graph.

LEMONADE SALES FOR ONE WEEK

Certain and Impossible

Lydia has a number cube with the numbers 1 through 6 on it.

Is rolling a number less than 7 a *certain* event or an *impossible* event?

To find out, list all the numbers Lydia could roll. You can then **predict,** or tell what will happen, when you roll the cube.

• If the event describes all the numbers, it is **certain.**

• If the event describes none of the numbers, it is **impossible.**

Numbers Lydia could roll: 1   2   3   4   5   6

Number less than 7?   <u>yes</u> <u>yes</u> <u>yes</u> <u>yes</u> <u>yes</u> <u>yes</u>

The event describes all the numbers. So, it is certain Lydia will roll a number less than 7.

For Problems 1–4, use the numbers below. Decide whether the event describes each number. Write *yes* or *no.*

1 2 3 4 5 6

1. a number greater than zero   ____ ____ ____ ____ ____ ____

2. a 2-digit number   ____ ____ ____ ____ ____ ____

3. a number less than 10   ____ ____ ____ ____ ____ ____

4. the number 7   ____ ____ ____ ____ ____ ____

Look at Problems 5–8. Decide whether each event is *certain* or *impossible.* Circle the correct answer.

5. a number greater than zero   certain   impossible

6. a 2-digit number   certain   impossible

7. a number less than 10   certain   impossible

8. the number 7   certain   impossible

Likely and Unlikely

Look at the spinner.

Is spinning a number with 2 as a digit a *likely* or an *unlikely* event?

To find out if an event is *likely* or *unlikely*, list all the numbers you could spin.

- If the event describes half, or more than half, of the numbers, it is *likely*.

- If the event describes fewer than half of the numbers, it is *unlikely*.

Numbers you could spin:	0	6	13	18	24	30
Number with 2 as a digit?	no	no	no	no	yes	no

The event describes fewer than half of the numbers. So, it is *unlikely* you will spin a number with 2 as a digit.

For Problems 1–4, use the spinner above. Decide whether the event describes each number. Write *yes* or *no*.

	0	6	13	18	24	30
1. a number greater than zero	____	____	____	____	____	____
2. a 2-digit number	____	____	____	____	____	____
3. a number less than 10	____	____	____	____	____	____
4. a number divisible by 3	____	____	____	____	____	____

For Problems 5–8, tell whether each event is *likely* or *unlikely*.

5. a number greater than zero _____

6. a 2-digit number _____

7. a number less than 10 _____

8. a number divisible by 3 _____

Predicting Outcomes

When you toss one coin, there are 2 possible ways the coin can land: heads or tails. Each way has the same chance of happening. The chances of heads or tails are equally likely.

When you toss two coins, there are 4 possible ways the coins can land.

There is 1 way to get 2 heads. →

There are 2 ways to get 1 head and 1 tail. ⌐→

There is 1 way to get 2 tails. →

POSSIBLE WAYS FOR TWO COINS TO LAND	
First Coin	Second Coin
heads	heads
heads	tails
tails	heads
tails	tails

So, tossing 1 head and 1 tail is a likely outcome.

When you spin once, there are 3 possible numbers you can get: 2, 3, or 4.

Complete the table to show all the possible sums you can get if you spin twice and add the numbers.

	POSSIBLE SUMS FOR TWO SPINS		
	First Spin	Second Spin	Sum
1.	2	2	2 + 2 = _____
2.	2	3	2 + 3 = _____
3.	2	_____	_____ + _____ = _____
4.	3	2	_____ + _____ = _____
5.	3	_____	_____ + _____ = _____
6.	3	_____	_____ + _____ = _____
7.	4	_____	_____ + _____ = _____
8.	4	_____	_____ + _____ = _____
9.	4	_____	_____ + _____ = _____

10. Look at the sums. Which sum is most likely to happen? _____

Probability

Probability can be described by a number written as a fraction.

$$\text{Probability} = \frac{\text{number of outcomes the event describes}}{\text{number of outcomes possible}}$$

Using the spinner at the right, what is the probability of spinning a circle?

- List all the outcomes possible.

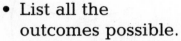 4 outcomes

- Identify the outcomes that are circles.

 1 circle

- Record the probability as a fraction.

$$\text{Probability circle} = \frac{\text{number of circles}}{\text{number of shapes}} = \frac{1}{4}$$

Use the spinner at the right. Circle the outcomes the event describes. Then write the probability as a fraction. The first one has been done for you.

1. What is the probability of spinning a ball?

 $\frac{3}{5}$

2. What is the probability of spinning a ball that is round?

 _____

3. What is the probability of spinning something that is not a ball?

 _____

4. What is the probability of spinning a football?

 _____

5. What is the probability of spinning a ball or a baseball bat?

 _____

Testing for Fairness

Fairness of a game is when one player is not more likely to win than another. To see if a game is fair, compare the probabilities of winning and not winning. A game is fair if the probabilities are equal.

<u>Game Description</u>

- One player tries to spin even numbers.
- One player tries to spin odd numbers.

Is the game fair if this spinner is used?

probability of spinning an even number $= \frac{3}{5}$
probability of spinning an odd number $= \frac{2}{5}$

The probabilities are not equal, so the game is not fair.

Use the spinner above. Write the probabilities. Write whether the game is *fair* or *not fair*.

<u>Game Description</u>

- One player tries to spin 9 or 12.

- One player tries to spin 8 or 12.

1. The probability of spinning 9 or 12 is ____.

2. The probability of spinning 8 or 12 is ____.

3. The game is _____.

Write the letter of the spinner that makes the game fair.

  A

  B

  C

4. Game Description

- One player tries to spin a rectangle.
- One player tries to spin a triangle.

Spinner _____

5. Game Description

- One player tries to spin a shape with 4 or fewer sides.
- One player tries to spin a shape with 4 or more sides.

Spinner _____

Name _____

Problem-Solving Strategy

Make a Model

Daryl, Tat, Judy, and Nora have written their first initials in sections of a spinner to play a game. The spinner has 8 equal sections. In 3 of the sections is a D. The sections with a T take up $\frac{1}{4}$ of the spinner. There is the same number of sections with J as with T. Each remaining section has an N. Is the spinner fair?

Make your model step by step, using one fact at a time.

• The spinner has 8 equal sections.	• In 3 of the sections is a D.	• The sections with a T take up $\frac{1}{4}$ of the spinner.

• There is the same number of sections with J as with T.	• Each remaining section has an N.	• Since Daryl has a greater chance of spinning his initial, D, the spinner is not fair.

Read the problem. Complete each fact and then complete the model to solve. Use crayons.

Kelsey made a spinner with 10 equal sections. She colored 2 opposite sections blue. She colored the section before each blue section orange. She made $\frac{2}{10}$ of the spinner green and $\frac{2}{10}$ of the spinner red. The rest of the sections she colored purple. Did Kelsey make a fair spinner?

1. How many sections are each color?

    A. blue_____   B. orange_____   C. green_____

    D. red_____   E. purple_____

2. The spinner _____ fair.

Exploring Geometric Figures

Geometric figures can be one-dimensional, two-dimensional, or three-dimensional. Figures such as lines, rectangles, and solids have **dimension** and can, therefore, be measured.

Lines, rays, and line segments are one-dimensional figures.	Circles, polygons, and planes are two-dimensional figures.	Solid figures are three-dimensional.

You can use linear units to measure the length (for example, feet, meters, inches, centimeters).	You can use square units to measure the area (for example, square inches, square meters).	You can use cubic units to measure the volume (for example, cubic inches, cubic yards).

For each example, write *one-dimensional, two-dimensional,* or *three-dimensional* to describe each figure.

1.

_____

2.

_____

3.

_____

4.

_____

5.

_____

6.

_____

Choose *feet, square feet,* or *cubic feet* to measure each.

7.

_____

8.

_____

9.

_____

Patterns for Solid Figures

You can make a three-dimensional figure
from a two-dimensional pattern called a **net.**
Look at the net at the right. Can you see how it
forms the three-dimensional rectangular prism?

Take another look at the net, and count the faces.

Since the three-dimensional rectangular prism has
6 faces, the net must also have 6 faces.

Draw a line from the three-dimensional figure to its net.

1.   2.   3.   4.

5.   6.   7.   8.

            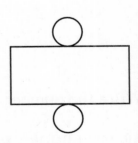

Faces of Solid Figures

A three-dimensional figure is made up of plane figures, known as faces.

A cube has 6 faces.

Cutting the cube apart gives you 6 plane figures, or 6 squares.

Some plane figures are triangles, rectangles, and squares.

Choose the plane figures that are the faces of each solid figure.

1.

2.

3.

a. rectangle

b. pentagon

c. triangle

d. square

_____   _____   _____

4.

5.

6.

_____   _____   _____

Write _triangles_, _rectangles_, or _squares_ for the plane figures that are the faces of each three-dimensional figure.

7.

8.

9.

_____   _____   _____

Write the number of faces for each three-dimensional figure.

10. rectangular prism _____   11.   square pyramid _____

More About Solid Figures

Three-dimensional figures are made up of faces.

- When two faces meet, they form a straight line, called an **edge.**

- A **vertex** is the point where three or more edges meet in a solid figure. **Vertices** is the word you use to describe more than one vertex.

Corner, or Vertex

Face

Edge

Circle each vertex and write the number of vertices.

1.

_____

2.

_____

3.

_____

Write the number of edges for each three-dimensional figure.

4.

_____

5.

_____

6.

_____

Write the number of faces for each three-dimensional figure.

7.

_____

8.

_____

9.

_____

Name _____

12.4

Plane Figures on a Coordinate Grid

You can find the point (3,4) on the grid.

Step 1 Start at 0.

Step 2 Go straight across to number 3.

Step 3 Go straight up to the line labeled 4.

The path to (3,4) is traced on the grid.

Always go straight **across** first,
and then go straight **up.**

Remember: Across begins with an **A,**
and **A** comes first in the alphabet.

Complete.

1. The numbers in an _____ represent
   the number of spaces you move across and up to
   locate a point on the grid.

Trace the path from zero to each of the following points on the
grid. Place a point.

2. (3,4)   3. (5,6)

4. (2,3)   5. (6,7)

6. (4,5)   7. (7,8)

8. Mark the ordered pairs on the grid.
   Place a point. Connect the points
   to make a figure.

   (2,3), (4,1), (7,1), (9,3), (7,5), (4,5)

9. Name the figure you drew. _____

R72 TAKE ANOTHER LOOK

Classifying and Sorting Solid Figures

There are many kinds of solid figures. They can have flat
faces, curved surfaces, or both flat faces and curved surfaces.

cylinder
2 flat faces
1 curved surface

cone
1 flat face
1 curved surface

cube
6 flat faces
0 curved surfaces

sphere
0 flat faces
1 curved surface

**rectangular
prism**
6 flat faces
0 curved surfaces

**triangular
prism**
5 flat faces
0 curved surfaces

**triangular
pyramid**
4 flat faces
0 curved surfaces

**square
pyramid**
5 flat faces
0 curved surfaces

Write the names of the solid figures that answer each question.

1. Which solid figures have both flat faces and curved surfaces?

_____

2. Which solid figures have a square face?

_____

3. Which solid figures have a curved surface?

_____

4. Which solid figures have no curved surfaces?

_____

_____

5. Which solid figures have fewer than 5 flat faces?

_____

Problem-Solving Strategy

Make an Organized List

You can classify and sort things in many ways. One strategy
is to *Make an Organized List.*

A school collected the following items for charities:
books, puzzles, paper towels, shampoo, cans of soup, boxes
of cornflakes, baseballs, oranges, notebooks, colored markers.

You can organize these items by geometric shape.

Box-Shaped	Sphere-Shaped	Cylinder-Shaped
books	baseballs	paper towels
puzzles	oranges	shampoo
boxes of cornflakes		cans of soup
notebooks		colored markers

Make an organized list to solve.

1. In Brigitte's locker there are books, coloring pens, a
   radio, a pencil, a lunch box, chalk, a tennis ball, and a
   pack of writing paper. Sort the objects by shape.

Box-Shaped	Cylinder-Shaped	Sphere-Shaped
_____	_____	_____
_____	_____	_____
_____	_____	_____
_____		_____

2. At the grocery store, Roger buys milk, butter, juice
   boxes, cereal, breadsticks, peanut butter, and an orange.
   Sort the objects by shape.

Box-Shaped	Cylinder-Shaped	Sphere-Shaped
_____	_____	_____
_____	_____	_____

Line Segments

Learning some basic terms can help you understand geometry.

Points identify locations on objects and in space. Points are named by letters. This is point *A*.

A
•

A **plane** is a flat surface with no end. Planes are named by any three points in the plane. This is plane *ABC*.

A **line** is a straight path in a plane. It has no end. It can be named by any two points on the line. This is line *AB*.

A **line segment** is the distance between two points on a line. This is line segment *AB*.

Parallel line segments never cross. They are always the same distance apart. Line segment *AB* is parallel to line segment *CD*.

Match the term with the drawing.

1. line ____   2. points ____   3. line segment ____   4. plane ____

A.

B.

C.

D. H •———————• I

Write *points*, *line*, or *line segment* to name each figure.

5. A •———————• B

6. •G   •H

7.

_____   _____   _____

Circle the parallel line segments.

8.

Name _____

LESSON
13.2

Exploring Angles and Line Relationships

A **ray** is part of a line and has one endpoint. You may think of the sun's rays, which start at the sun (endpoint) and go on forever into space.

A————B
ray AB

An **angle** is formed when two rays have the same endpoint.

There are three types of angles:

A **right angle** forms a square corner.

An **acute angle** is less than a right angle.

An **obtuse angle** is greater than a right angle.

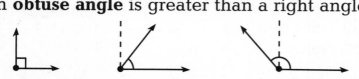

angle C
of *ECD*

right angle   acute angle   obtuse angle

Hint: *Obtuse* begins with an **O**. Think of an **O**pen angle.

Write the name of each figure.

1.
E   F

2.
Y   Z

3.
B

4.

_____   _____   _____

Write *acute, right,* or *obtuse* to name each angle.

5.

6.

7.

_____   _____   _____

8.

9.

10.

_____   _____   _____

R76   **TAKE ANOTHER LOOK**

More About Angles and Line Relationships

All lines that cross each other are **intersecting lines.**
Lines *DE* and *FG* intersect.

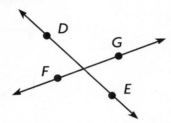

Lines that intersect to form four right angles are
perpendicular lines. Lines *AB* and *CD* are
perpendicular lines.

In Exercises 1–3, draw perpendicular lines.

1.

2.

3.

For Exercises 4–5, use the following map.

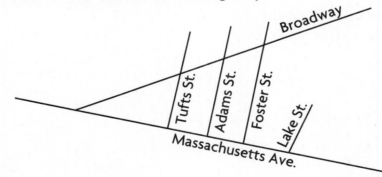

4. Name the streets parallel to Adams Street. _____

5. Name the streets perpendicular to Massachusetts Avenue.

_____

Exploring Circles

A **circle** is named by its center point. All the points on the circle are the same distance from its center point.

A **radius** of a circle is a line segment from the center to any point on the circle.

A **diameter** of a circle is a line segment that passes through the center and has its endpoints on the circle.

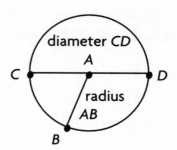

For Problems 1–4, use Circle 1 and a centimeter ruler.

1. The center of the circle is point _____.

2. Name each radius of the circle.

   _____

   Each radius measures _____ cm.

3. The diameter of the circle is line segment

   _____.

   The diameter measures _____ cm.

4. Name three points on the circle. _____

Circle 1

5. Draw a circle. Label the center point *R*.
   Draw a radius *RS*. Draw a diameter *TV*.

   Measure the radius. _____

   Measure the diameter. _____

For Problems 6–8, use Circle 2.

6. Name the center of the circle. _____

7. Name each radius of the circle. _____

8. Name the diameter of the circle. _____

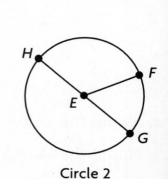

Circle 2

Polygons

A **polygon** is a closed plane figure with straight sides.
A polygon is named by its number of sides or angles.

A **triangle** has three sides and three angles. The prefix
tri- has three letters.

A **quadrilateral** has four sides and four angles. The prefix
quad- has four letters.

A **pentagon** has five sides and five angles. The prefix *penta-*
has five letters.

A **hexagon** has six sides and six angles. The prefix
hex- has an x like *six*.

An **octagon** has eight sides and eight angles. Remember
that it is similar to an eight-armed *oct*opus.

Write *yes* or *no* to tell if the figure is a polygon.

1. _____   2. _____   3. _____   4. _____

5. _____   6. _____   7. _____   8. _____

Name each polygon. Tell how many sides and angles.

9.   10.   11.   12.

_____   _____   _____   _____

_____   _____   _____   _____

_____   _____   _____   _____

Use a straightedge to draw each figure.

13. pentagon   14. hexagon

Quadrilaterals

A **quadrilateral** has four sides and four angles.
Quadrilaterals are grouped by their angles, by their sides,
and by any parallel sides. The table will help you compare
different quadrilaterals.

Grouping	Side Lengths	Parallel Sides	Angles
General Quadrilaterals	4 different lengths	sides may or may not be parallel	angles vary
Trapezoids	lengths can vary	1 pair parallel sides	angles vary
Parallelograms	2 pairs equal lengths	2 pairs parallel sides	2 acute, 2 obtuse, or 4 right
Rectangles	2 pairs equal lengths	2 pairs parallel sides	4 right
Rhombuses	4 equal lengths	2 pairs parallel sides	2 acute, 2 obtuse
Squares	4 equal lengths	2 pairs parallel sides	4 right

Match the drawing with the term.

1. ____

2. ____

3. ____

4. ____

5. ____

6. ____

A. rhombus   **B.** trapezoid

C. square   **D.** rectangle

E. parallelogram

F. general quadrilateral

For Exercises 7–8, use the table above.

7. How are rhombuses and squares alike?

_____

8. How are rhombuses and squares different?

_____

_____

Name _____

Problem-Solving Strategy

Act It Out

You can solve a problem involving geometric figures by *acting it out*.

Trace the triangles at the right and cut them out.

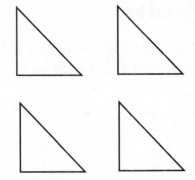

For Exercises 1–4, use the four triangles to make different quadrilaterals. Draw a picture of each quadrilateral.

1. Draw a square.

2. Draw a rectangle.

3. Draw a trapezoid.

4. Draw a parallelogram.

Solve.

5. Lockers are grouped in fours at the Melick School. Ariana's locker is after Hasan's but before George's. Eraldo's locker is after Hasan's but before Ariana's. If Ariana's locker is next to George's, in what order are the lockers?

_____

_____

6. The school store sold T-shirts at the school fair. In the morning 134 shirts were sold. In the afternoon 187 shirts were sold. The store had 29 shirts remaining. How many shirts did the store have at the beginning of the day?

_____

Finding Perimeter

Perimeter is the distance around a figure. You can find the perimeter by adding the lengths of the sides of a figure.

The perimeter is
6 + 4 + 6 + 4 = 20 units.

The perimeter is
1 in. + 3 in. + 2 in. + 7 in. + 3 in. + 10 in. = 26 in.

If two pairs of sides are the same length, you can find the perimeter by multiplying and adding.

$(2 \times 3 \text{ ft}) + (2 \times 9 \text{ ft}) =$
6 ft + 18 ft = 24 ft

$(2 \times 2 \text{ in.}) + (2 \times 10 \text{ in.}) =$
4 in. + 20 in. = 24 in.

Find the perimeter.

1.

_____

2.

_____

3.

_____

4.

_____

5.

_____

6.

_____

7.

_____

8. 10 in.
3in. [   ] 3 in.
10 in.

_____

Exploring Area

Area is the number of square units needed to cover a flat surface.

You can find the area of each figure on the geoboard by counting the number of square units.

The area is
6 square units.

$\frac{1}{2}$ unit

$\frac{1}{2}$ unit

The area is
2 square units.

The area is
8 square units.

Find the area.

1.   _____

2.   _____

3.   _____

4.   _____

5.   _____

6.   _____

7.   _____

8.   _____

9.   _____

10.   _____

11.   _____

12.   _____

13. Which figures have the same area? _____

a.

b.

c.

d.

Area of Irregular Figures

Here is one way to find the area of the figure.

Step 1   Start by counting the whole square units. There are 14 whole square units.

Step 2   Then count the partial square units. In this figure there are 6 half units, which are equal to 3 whole units.

Step 3   Add together the whole and partial square units. 14 + 3 = 17

So, the figure's area is 17 square units.

Find the area.

1.

   _____

2.

   _____

3.

   _____

4.

   _____

5.

   _____

6.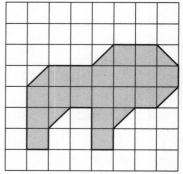

   _____

Finding Area

You can find the area of a rectangle by using multiplication.

You multiply the length times the width of a rectangle to find the area.

length × width = area

7 ft × 3 ft = 21 sq ft

7 ft
3 ft

Write your answer in square units, such as *sq ft* or *sq yd*.

Find the area.

1.
6 in.
3 in.

_____

2.

4 yd
4 yd

_____

3.
3 km
8 km

_____

4.
7 ft
5 ft

_____

5.

9 cm
8 cm

_____

6.

8 ft
1 ft

_____

Circle the letter of the rectangle that has the greater area.

7. a.
6 ft
7 ft

b.
7 ft
5 ft

8. a.

9 in.
5 in.

b.
8 in.
7 in.

9. a. 5 km

5 km

b.

9 km
3 km

10. a. 6 cm

3 cm

b.

8 cm
2 cm

Relating Area and Perimeter

Perimeter is the distance around a figure. Perimeter is measured in linear units.

Area is the number of square units needed to cover a flat surface. Area is measured in square units.

Figures with the same area can have different perimeters.

Area = 20 sq units
Perimeter = 22 units

Area = 20 sq units
Perimeter = 24 units

Figures with the same perimeter can have different areas.

Area = 16 sq units
Perimeter = 28 units

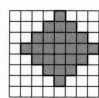

Area = 25 sq units
Perimeter = 28 units

Write the area and the perimeter.

1.

2.

3.

_____   _____   _____

_____   _____   _____

For each figure, draw another figure that has the same area but a different perimeter. Write the area in square units.

4.

5.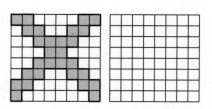

area = _____   area = _____

Problem-Solving Strategy

Draw a Diagram

You can draw a diagram to solve problems involving area and perimeter. If you know the perimeter, you can figure out the greatest area.

Jane has 28 feet of fencing to go around a rectangular vegetable garden. She wants to know the shape of the rectangle that will give her the greatest area. She can draw a diagram of each shape that has a perimeter of 28 feet.

perimeter = 28 ft
area = 13 sq ft

perimeter = 28 ft
area = 24 sq ft

perimeter = 28 ft
area = 33 sq ft

perimeter = 28 ft
area = 40 sq ft

perimeter = 28 ft
area = 45 sq ft

perimeter = 28 ft
area = 49 sq ft

So, a square 7 feet by 7 feet has the greatest area.

Draw a diagram to solve.

1. Tad has 16 feet of fencing to put around his garden. What is the greatest area he can make his garden?

_____

2. George has 36 feet of fencing to build a pen for his duck. What shape will give him the greatest area? What are the lengths of the sides?

_____

Use the space below to draw a diagram to solve Problems 1–2.

Translations, Reflections, and Rotations

When you have moved a figure, you have completed a **transformation**.
Three kinds of transformations are described below.

- Sliding a figure in any direction is called a **translation.** The figure looks exactly the same but is in a new position.

- You can flip a figure over an imaginary line. The flipped figure is a mirror image called a **reflection.**

- A figure that is turned around a point is called a **rotation.** The point holds the corner of the figure while the rest of the figure turns around the point.

Tell how each figure was moved. Write *translation, reflection,* or *rotation.*

1.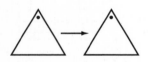

2.

3.

4.

_____ _____ _____ _____

Copy each original figure. Draw figures to show a *translation, reflection,* and *rotation.*

5.   Original   Translation   Reflection   Rotation

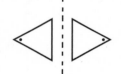

6.

Congruence

Congruent figures have the same size and shape.

The figures at the right are congruent.
They are not in the same position, but
they are the same size and shape.

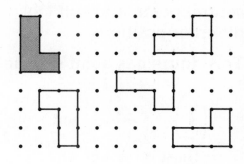

You can check for congruency by tracing
one figure on another sheet of paper. Place
the traced figure over the other figures and
compare for congruency.

Trace one figure from each set. Place it over the second figure and
check for congruency. Write whether each pair is *congruent* or
not congruent.

1.

_____

2.

_____

3.

_____

4.

_____

5.

_____

6.

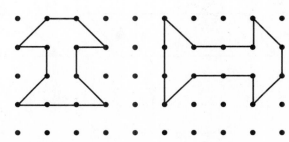

_____

Two Kinds of Symmetry

A figure has **point symmetry** if it can be turned about a central point, and still look the same in at least two different positions.

This figure has point symmetry.

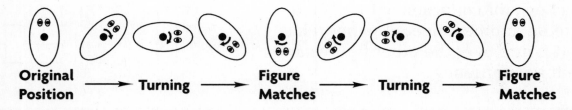

Original Position → **Turning** → **Figure Matches** → **Turning** → **Figure Matches**

This figure *does not* have point symmetry.

Original Position → **Turning** → **No Matches** → **Turning** → **Figure Matches**

Trace the following figures, and turn them to determine whether each figure has point symmetry. Write *yes* or *no.*

1.

_____

2.

_____

3.

_____

4.

_____

5.

_____

6.
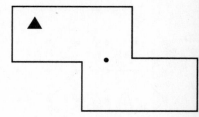

_____

More About Symmetry

Trace and cut out the figure on the right. Fold it along the dotted line.

When you unfold the figure, you can see that one side is a reflection of the other. When a figure can be folded about a line so that its two parts are identical, the figure has **line symmetry**.

Is the dotted line a line of symmetry? Trace and cut out each figure. Fold to check. Write *yes* or *no*.

1.

_____

2.

_____

3.

_____

4.

_____

5.

_____

6.

_____

Draw the other half of the figure to show that the figure has line symmetry.

7.

8.

9.

Tessellations

When you arrange polygons to cover a surface without leaving any space between them or making them overlap, you are making a **tessellation**.

Here are some tessellations:

Squares and Octagons   Hexagons   Right Triangles

The polygons in these tessellations all connect with no open spaces or overlaps.

Use two different colors to make your own tessellation. Use the polygons given.

1.

2.

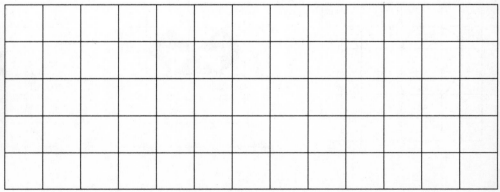

Name _____

LESSON
15.5

Changing Sizes of Shapes

Similar figures do not have to be the same size, but they
have to be the same shape. To change the size of a figure,
follow these steps.

Step 1
Place corner points
on the figure.
Count the spaces
between each point.

Step 2
Count the same number
of spaces on a smaller
or a larger grid. Place
the points.

Step 3
Complete the sides of
the similar figure.

Reduce each figure by using the grid below.

1.

2.

3.

TAKE ANOTHER LOOK   R93

Problem-Solving Strategy

Make a Model

When you need to solve a problem involving geometric figures, it may be helpful to make a model.

Fred is building a fence 10 yards long. He will place a post at each end of the fence and posts every 2 yards between the ends. How many fence posts does he need?

Step 1 Draw a straight line to represent the fence.

Step 2 Use models to represent the fence posts. Move the models around to match the problem.

Step 3 Look back at the problem. If your model matches, answer the question.

Fred needs 6 posts.

Make a model to solve.

1. T.J. is putting up a 12-foot shelf in the garage. He will hang the shelf with a bracket at each end of the shelf and brackets every 3 feet between the ends. How many brackets does he need?

   _____

2. Michael has $1.10 in coins in his pocket. He has 5 dimes and twice as many nickels as dimes. The rest of his coins are pennies. How many of each coin does he have?

   _____

3. Five students stood in a line for water. Holly was in the middle. Bob stood next to Holly. Chris was at the end. Anne stood between Holly and Chris. Mark was at the fountain. Show the students' order.

   _____

Multiplying by Multiples

A **multiple** is the product of a given number and another whole number. For example, 60 is a multiple of 6 and 10: $6 \times 10 = 60$. You can use basic multiplication facts and mental math to find the product.

Solve $60 \times 2 = \underline{}$.

Step 1 Think of the basic fact. Multiply. $6 \times 2 = 12$

Step 2 Count the number of zeros in the factors. $6\underline{0}$ has 1 zero.

Step 3 Add the number of zeros in the factors   The answer is $12\underline{0}$.
to the basic-fact product.

Write the basic multiplication fact.

1. $70 \times 4 = 280$ **2.** $2{,}000 \times 2 = 4{,}000$ **3.** $600 \times 8 = 4{,}800$

_____   _____   _____

Finish each pattern.

4. $3 \times 3 = \underline{\hphantom{xxxx}}$ **5.** $7 \times 7 = \underline{\hphantom{xxxx}}$ **6.** $5 \times 4 = \underline{\hphantom{xxxx}}$

$30 \times 3 = \underline{\hphantom{xxxx}}$ $70 \times 7 = \underline{\hphantom{xxxxx}}$ $50 \times 4 = \underline{\hphantom{xxxxx}}$

$300 \times 3 = \underline{\hphantom{xxxxx}}$ $700 \times 7 = \underline{\hphantom{xxxxx}}$ $500 \times 4 = \underline{\hphantom{xxxxxx}}$

$3{,}000 \times 3 = \underline{\hphantom{xxxxxx}}$ $7{,}000 \times 7 = \underline{\hphantom{xxxxxx}}$ $5{,}000 \times 4 = \underline{\hphantom{xxxxx}}$

7. $6 \times 2 = \underline{\hphantom{xxxx}}$ **8.** $4 \times 8 = \underline{\hphantom{xxxx}}$ **9.** $3 \times 9 = \underline{\hphantom{xxxx}}$

$60 \times 2 = \underline{\hphantom{xxxxx}}$ $40 \times 8 = \underline{\hphantom{xxxxx}}$ $30 \times 9 = \underline{\hphantom{xxxxx}}$

$600 \times 2 = \underline{\hphantom{xxxxx}}$ $400 \times 8 = \underline{\hphantom{xxxxx}}$ $300 \times 9 = \underline{\hphantom{xxxxx}}$

$6{,}000 \times 2 = \underline{\hphantom{xxxxx}}$ $4{,}000 \times 8 = \underline{\hphantom{xxxxx}}$ $3{,}000 \times 9 = \underline{\hphantom{xxxxx}}$

10. $3 \times \underline{\hphantom{xxxx}} = 18$ **11.** $7 \times 5 = \underline{\hphantom{xxxx}}$ **12.** $9 \times \underline{\hphantom{xxxx}} = 36$

$\underline{\hphantom{xxxx}} \times 6 = 180$ $\underline{\hphantom{xxxx}} \times 5 = 350$ $\underline{\hphantom{xxxx}} \times 4 = 360$

$\underline{\hphantom{xxxxx}} \times 6 = 1{,}800$ $700 \times 5 = \underline{\hphantom{xxxxx}}$ $900 \times 4 = \underline{\hphantom{xxxxx}}$

$3{,}000 \times 6 = \underline{\hphantom{xxxxx}}$ $7{,}000 \times \underline{\hphantom{xxxx}} = 35{,}000$ $\underline{\hphantom{xxxxx}} \times 4 = 36{,}000$

A Way to Multiply

Suppose you have 3 boxes of golf balls. Each box contains 25 balls. How many golf balls do you have?

You can solve the problem by using base-ten blocks.

After regrouping the ones, you have 7 tens and 5 ones, or 75. So, you have 75 golf balls.

You can also solve the problem by using a repeated addition number sentence. You know you have 3 groups of 25. So, $25 + 25 + 25 = 75$, or $3 \times 25 = 75$.

Write a repeated addition number sentence for each problem. Then write a multiplication number sentence.

1. Paul's Pizzeria ordered 7 boxes of cheese. Each box contained 50 pounds of cheese. How much did the 7 boxes of cheese weigh?

   _____

   _____

2. In Mr. Palmer's class, 6 students each read 26 books for a reading contest. How many books in all did the 6 students read?

   _____

   _____

3. The 10 members of the school's track club ran a 6-mile race. How many miles in all did the 10 members run?

   _____

   _____

4. A farm donated 9 boxes of fruit to a local food bank. Each box contained 48 pieces of fruit. How many pieces of fruit did the farm donate in all?

   _____

   _____

Find the product.

5. $\begin{array}{r} 37 \\ \times\ 6 \\ \hline \end{array}$
6. $\begin{array}{r} 69 \\ \times\ 4 \\ \hline \end{array}$
7. $\begin{array}{r} 42 \\ \times\ 4 \\ \hline \end{array}$
8. $\begin{array}{r} 27 \\ \times\ 8 \\ \hline \end{array}$

Modeling Multiplication

Solve. $17 \times 6 = \underline{?}$, or $\begin{array}{r} 17 \\ \times\,6 \\ \hline \end{array}$

- You can find the product by breaking the two-digit number apart.

- Multiply, using the simpler numbers.

First, multiply 6 times 7. $6 \times 7 = 42$ ⎫

Second, multiply 6 times 10. $6 \times 10 = 60$ ⎬ partial products

Third, add the **partial products** to find the product.
$42 + 60 = 102$

Find each product by using partial products. Show the three steps you used.

1. $\begin{array}{r} 41 \\ \times\,6 \\ \hline \end{array}$

   $6 \times 1 = 6$ ___

   $6 \times 40 = 240$ ___

   $6 + 240 = 246$ ___

2. $\begin{array}{r} 37 \\ \times\,3 \\ \hline \end{array}$

   _____

   _____

   _____

3. $\begin{array}{r} 67 \\ \times\,2 \\ \hline \end{array}$

   _____

   _____

   _____

4. $\begin{array}{r} 55 \\ \times\,4 \\ \hline \end{array}$

   _____

   _____

   _____

5. $\begin{array}{r} 29 \\ \times\,7 \\ \hline \end{array}$

   _____

   _____

   _____

6. $\begin{array}{r} 32 \\ \times\,5 \\ \hline \end{array}$

   _____

   _____

   _____

7. $\begin{array}{r} 46 \\ \times\,6 \\ \hline \end{array}$

   _____

   _____

   _____

8. $\begin{array}{r} 71 \\ \times\,3 \\ \hline \end{array}$

   _____

   _____

   _____

9. $\begin{array}{r} 54 \\ \times\,8 \\ \hline \end{array}$

   _____

   _____

   _____

Recording Multiplication

Find the product 2 × 434.

Step 1	Step 2	Step 3
Multiply the ones.	Multiply the tens.	Multiply the hundreds.

H	T	O
4	3	**4**
×		**2**
		8

H	T	O
4	**3**	4
×		**2**
	6	8

H	T	O
4	3	4
×		**2**
8	6	8

So, 2 × 434 = 868.

Find the product 4 × 367. You may need to regroup.

Step 1	Step 2	Step 3
Multiply the ones. Regroup?	Multiply the tens. Add the regrouped tens. Regroup?	Multiply the hundreds. Add the regrouped hundreds.

H	T	O
3	2 6	**7**
×		**4**
		8

Regroup 28

H	T	O
2 3	2 **6**	7
×		**4**
	6	8

Regroup 26

H	T	O
2 **3**	2 6	7
×		**4**
14	6	8

So, 4 × 367 = 1,468.

Complete.

1.

H	T	O
1	2	3
×		3

2.

H	T	O
3	1	1
×		4

3.

H	T	O
4	2	7
×		6

Practicing Multiplication

Use estimation when you multiply with problems involving money.

Find the product 4 × $2.99.

2	9	9
×		4
$11	9	6

- First, estimate your answer. 4 × $3.00 is $12.00.

- Second, multiply by using whole numbers.

- Third, compare your answer with your estimate. Use your estimate to help you place the decimal point and dollar sign.

- Since 1196 is close to 1200, you place the decimal point between the 1 and the 9. Add the dollar sign.

Estimate the product. Then place the dollar sign and decimal point in each given product.

1. $6.62
      ×   6 _____
      3 9 7 2

2. $9.34
      ×   3 _____
      2 8 0 2

3. $7.54
      ×   7 _____
      5 2 7 8

Multiply and record each product.

4.
$	1.	7	4
×			5
$	.		

5.
$	2.	2	3
×			7
$	.		

6.
$	4.	9	2
×			6
$	.		

7.
$	3.	3	7
×			4
$	.		

8.
$	2.	4	9
×			8
$	.		

9.
$	3.	8	7
×			3
$	.		

10.
$	1.	9	8
×			5
$	.		

11.
$	4.	2	1
×			6
$	.		

12.
$	7.	1	3
×			4
$	.		

Problem-Solving Strategy

Write a Number Sentence

Writing a number sentence helps organize the information within a math problem. Consider the following word problem. Bill wants to buy 7 sets of collectors' cards. Each set cost $2.99. How much money does Bill need?

number of sets	×	cost per set	=	total cost
7	×	$2.99	=	n

$7 \times \$2.99 = n$ $\qquad$ $n = \$20.93$

Write a number sentence to solve the problem.

1. Leslie buys 5 books. The books cost $3.99 each. How much does her purchase cost in all?

   _____

2. Mark exercises for 50 minutes a day, 5 days a week. How many minutes each week does Mark exercise?

   _____

3. Last year, Lorna's mother traveled 7,496 miles on business trips. This year, she traveled 6,957 miles. How many more miles did Lorna's mother travel last year?

   _____

4. Emma paid $39.95 for a CD, $19.95 for a user's manual, and $79.95 for a game stick. How much did she spend in all?

   _____

5. Joy practices the piano for 35 minutes each day. How many minutes does she practice in one week?

   _____

Patterns with Multiples

If you can multiply ones,
then you can multiply tens.

$$\begin{array}{r} 2 \\ \times\,3 \\ \hline 6 \end{array} \qquad \begin{array}{r} 2\text{ tens} \\ \times\,3 \\ \hline 6\text{ tens} \end{array} \qquad = \begin{array}{r} 20 \\ \times\,3 \\ \hline 60 \end{array}$$

You can also multiply
hundreds.

$$\begin{array}{r} 4 \\ \times\,2 \\ \hline 8 \end{array} \qquad \begin{array}{r} 4\text{ hundreds} \\ \times\,2 \\ \hline 8\text{ hundreds} \end{array} \qquad = \begin{array}{r} 400 \\ \times\,2 \\ \hline 800 \end{array}$$

You can even multiply
thousands!

$$\begin{array}{r} 6 \\ \times\,6 \\ \hline 36 \end{array} \qquad \begin{array}{r} 6\text{ thousands} \\ \times\,6 \\ \hline 36\text{ thousands} \end{array} \qquad = \begin{array}{r} 6{,}000 \\ \times\,6 \\ \hline 36{,}000 \end{array}$$

Complete.

1. $\begin{array}{r} 6 \\ \times\,4 \\ \hline \end{array}$ $\begin{array}{r} 6\text{ tens} \\ \times\,4 \\ \hline \end{array}$ $\begin{array}{r} 60 \\ \times\,4 \\ \hline \end{array}$ $\begin{array}{r} 6\text{ hundreds} \\ \times\,4 \\ \hline \end{array}$ $\begin{array}{r} 600 \\ \times\,4 \\ \hline \end{array}$ $\begin{array}{r} 6\text{ thousands} \\ \times\,4 \\ \hline \end{array}$ $\begin{array}{r} 6{,}000 \\ \times\,4 \\ \hline \end{array}$

2. $\begin{array}{r} 7 \\ \times\,6 \\ \hline \end{array}$ $\begin{array}{r} 7\text{ tens} \\ \times\,6 \\ \hline \end{array}$ $\begin{array}{r} 70 \\ \times\,6 \\ \hline \end{array}$ $\begin{array}{r} 7\text{ hundreds} \\ \times\,6 \\ \hline \end{array}$ $\begin{array}{r} 700 \\ \times\,6 \\ \hline \end{array}$ $\begin{array}{r} 7\text{ thousands} \\ \times\,6 \\ \hline \end{array}$ $\begin{array}{r} 7{,}000 \\ \times\,6 \\ \hline \end{array}$

3. $\begin{array}{r} 8 \\ \times\,5 \\ \hline \end{array}$ $\begin{array}{r} 8\text{ tens} \\ \times\,5 \\ \hline \end{array}$ $\begin{array}{r} 80 \\ \times\,5 \\ \hline \end{array}$ $\begin{array}{r} 8\text{ hundreds} \\ \times\,5 \\ \hline \end{array}$ $\begin{array}{r} 800 \\ \times\,5 \\ \hline \end{array}$ $\begin{array}{r} 8\text{ thousands} \\ \times\,5 \\ \hline \end{array}$ $\begin{array}{r} 8{,}000 \\ \times\,5 \\ \hline \end{array}$

4. $\begin{array}{r} 9 \\ \times\,6 \\ \hline \end{array}$ $\begin{array}{r} 9\text{ tens} \\ \times\,6 \\ \hline \end{array}$ $\begin{array}{r} 90 \\ \times\,6 \\ \hline \end{array}$ $\begin{array}{r} 9\text{ hundreds} \\ \times\,6 \\ \hline \end{array}$ $\begin{array}{r} 900 \\ \times\,6 \\ \hline \end{array}$ $\begin{array}{r} 9\text{ thousands} \\ \times\,6 \\ \hline \end{array}$ $\begin{array}{r} 9{,}000 \\ \times\,6 \\ \hline \end{array}$

5. $\begin{array}{r} 4 \\ \times\,8 \\ \hline \end{array}$ $\begin{array}{r} 4\text{ tens} \\ \times\,8 \\ \hline \end{array}$ $\begin{array}{r} 40 \\ \times\,8 \\ \hline \end{array}$ $\begin{array}{r} 4\text{ hundreds} \\ \times\,8 \\ \hline \end{array}$ $\begin{array}{r} 400 \\ \times\,8 \\ \hline \end{array}$ $\begin{array}{r} 4\text{ thousands} \\ \times\,8 \\ \hline \end{array}$ $\begin{array}{r} 4{,}000 \\ \times\,8 \\ \hline \end{array}$

6. $\begin{array}{r} 5 \\ \times\,7 \\ \hline \end{array}$ $\begin{array}{r} 5\text{ tens} \\ \times\,7 \\ \hline \end{array}$ $\begin{array}{r} 50 \\ \times\,7 \\ \hline \end{array}$ $\begin{array}{r} 5\text{ hundreds} \\ \times\,7 \\ \hline \end{array}$ $\begin{array}{r} 500 \\ \times\,7 \\ \hline \end{array}$ $\begin{array}{r} 5\text{ thousands} \\ \times\,7 \\ \hline \end{array}$ $\begin{array}{r} 5{,}000 \\ \times\,7 \\ \hline \end{array}$

Problem-Solving Strategy

Find a Pattern

Carrie has finished part of her square quilt. How many patches of each color will be in the final quilt?

Look for patterns to help you make a plan for solving.

Since this is a square quilt and there are 9 patches along the top, you know there must be 9 rows.

Look closely at the quilt. There is a pattern of 3 blues, 3 reds, and 3 whites in each row. Multiply each color by the 9 rows. So, there are 27 blue, 27 red, and 27 white patches.

B	R	W	B	R	W	B	R	W
R	W	B	R	W	B	R	W	B
B	R	W	B	R	W	B	R	W
R	W	B	R	W	B	R	W	B
B	R	W	B	R	W	B	R	W
R	W	B	R	W	B	R	W	B

B = Blue   R = Red   W = White

Blue	Red	White
$3 \times 9 = 27$	$3 \times 9 = 27$	$3 \times 9 = 27$

Find a pattern and solve.

1. Shade the figure to finish the pattern.

2. There are an equal number of trucks and cars in the lot. Each truck has 8 wheels, and each car has 4 wheels. How many trucks and cars are there if there are 36 wheels?

_____

3. The Kopi Center is open from 9 A.M. to 7 P.M., 7 days a week. How many hours is it open in a 30-day month?

_____

Estimating Products with Multiples

When you don't need an exact answer, you can estimate by rounding the factors.

To round:

- Circle the digit to be rounded.

- Underline the digit to its right.

- If the underlined digit is 5 or greater, increase the circled digit by 1. Otherwise, the circled digit stays the same.

- Write zeros in all places to the right of the circled digit.

⑤5̲1 →   600 ← rounded up
× ②4̲ →   × 20 ← rounded down
                    12,000

Rewrite each problem by rounding each factor. Then estimate the product.

1.   62 →   60
   × 24 → × 20
              1,200

2.   43 → _____
   × 37 → × _____

   _____

3.   75 → _____
   × 24 → × _____

   _____

4.   28 → _____
   × 17 → × _____

   _____

5.   44 → _____
   × 23 → × _____

   _____

6.   94 → _____
   × 36 → × _____

   _____

7. $37 → _____
   × 34 → × _____

   _____

8.   55 → _____
   × 27 → × _____

   _____

9. $84 → _____
   × 38 → × _____

   _____

10. 448 → _____
   × 57 → × _____

   _____

11. 637 → _____
   × 42 → × _____

   _____

12. $275 → _____
   × 75 → × _____

   _____

Modeling Multiplication

A model can be used to simplify multiplication.

This is a model for 12×26.

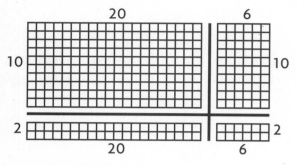

Multiply to find the number of squares in each rectangle. Then add the partial products.

$2 \times 6 = \qquad 12$

$2 \times 20 = \qquad 40$

$10 \times 6 = \qquad 60$

$\underline{10 \times 20 = +\ 200}$

$\mathbf{312}$

When you add the value of each rectangle in the grid, you find that the product of 12 and 26 is 312.

Multiply and solve by using the model and writing partial products.

1. $\begin{array}{r} 26 \\ \times 14 \\ \hline ? \end{array}$

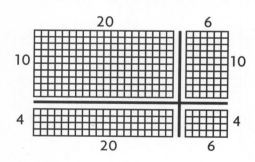

$4 \times 6 = \underline{\qquad}$

$4 \times 20 = \underline{\qquad}$

$10 \times 6 = \underline{\qquad}$

$\underline{10 \times 20 = +}$

$\underline{\qquad}$

2. $\begin{array}{r} 23 \\ \times 17 \\ \hline ? \end{array}$

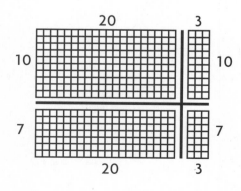

$7 \times 3 = \underline{\qquad}$

$7 \times 20 = \underline{\qquad}$

$10 \times 3 = \underline{\qquad}$

$\underline{10 \times 20 = +}$

$\underline{\qquad}$

Recording Multiplication

Solve 12×15 by using the partial-products method.

Step 1   **Step 2**   **Step 3**   **Step 4**

$$\begin{array}{r} 1\,5 \\ \times\, 1\,2 \end{array}$$

Step 5

Add partial products together.

$10 + 20 + 50 + 100 = 180$

The product of 12 and 15 is 180.

Partial Products

		1	5
	$\times$	1	2
$2 \times 5 =$		1	0
$2 \times 10 =$		2	0
$10 \times 5 =$		5	0
$10 \times 10 =$	1	0	0
product $\rightarrow$	1	8	0

Multiply by using the partial-products method.

1.

$$\begin{array}{r} 3\ \ 1 \\ \times\ 1\ \ 7 \end{array}$$

$7 \times 1 =$ _____

$7 \times 30 =$ _____

$10 \times 1 =$ _____

$10 \times 30 =$ _____

product $\rightarrow$ _____

2.

$$\begin{array}{r} 4\ \ 6 \\ \times\ 2\ \ 8 \end{array}$$

$8 \times 6 =$ _____

$8 \times 40 =$ _____

$20 \times 6 =$ _____

$20 \times 40 =$ _____

product $\rightarrow$ _____

3.

$$\begin{array}{r} 7\ \ 9 \\ \times\ 5\ \ 6 \end{array}$$

$6 \times 9 =$ _____

$6 \times 70 =$ _____

$50 \times 9 =$ _____

$50 \times 70 =$ _____

product $\rightarrow$ _____

4.
$$\begin{array}{r} 82 \\ \times 25 \end{array}$$

5.
$$\begin{array}{r} 63 \\ \times 47 \end{array}$$

6.
$$\begin{array}{r} 92 \\ \times 34 \end{array}$$

Practicing Multiplication

Find the product 28 × 136.

- Estimate by rounding. Since 30 × 100 is close to 3,000, your answer should be close to 3,000.

- Multiply.

```
  136
×  28
1,088  ← 136 × 8
2,720  ← 136 × 20
3,808
```

- Compare the product with your estimate. Since 3,808 is close to your estimate of 3,000, it is a reasonable product.

Remember:

136 × 8 is the same as

```
8 × 6   =      48
8 × 30  =     240
8 × 100 =   +800
            1,088
```

Estimate and multiply.

1. 126 → 100
   × 45 → × 50
   ‾‾‾‾‾‾‾‾‾‾‾
          5,000

		1	2	6
×			4	5
		6	3	0
+	5,	0	4	0
	5,	6	7	0

2. 276 → _____
   × 38 → × _____
   ‾‾‾‾‾‾‾‾‾‾‾‾‾‾‾
   _____

		2	7	6
×			3	8
+				

3. 737 → _____
   × 47 → × _____
   ‾‾‾‾‾‾‾‾‾‾‾‾‾‾‾

4. 545 → _____
   × 24 → × _____
   ‾‾‾‾‾‾‾‾‾‾‾‾‾‾‾

Modeling Division

Divide 92 into 4 equal groups.

Write. 92 ÷ 4

Step 1

Show 92 as 9 tens and 2 ones.

Draw 4 circles to show 4 groups.

○ ○ ○ ○

Step 2

Begin by dividing the 9 tens. Place an equal number of tens into each circle.

2 — 2 tens in each group
−8 — 8 tens used
1 — 1 ten left

Step 3

Regroup the 1 ten left over into ones. Now there are 12 ones.

bring ones down

Step 4

Divide the 12 ones. Place an equal number of ones into each circle.

23 — 3 ones in each group
4)92
−8
12
−12 — 12 ones used
0 — 0 ones left

---

92 ÷ 4 = 23

Use base-ten blocks to model each problem. Record the numbers
as you complete each step.

1. 2)4 8   2. 3)9 6   3. 4)5 2   4. 2)3 4   5. 5)6 5

Dividing with Remainders

What is $23 \div 5$?

The drawings in the boxes below show two different ways to think about this division problem.

Example A

Think: If you divide 23 into 5 equal groups, how many are in each group?

$5 \times \boxed{?} = 23$

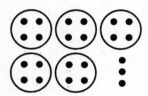

There are 5 groups of 4, with 3 left over.

Example B

Think: How many groups of 5 can you make from 23?

$\boxed{?} \times 5 = 23$

There are 4 groups of 5, with 3 left over.

$23 \div 5 = 4 \text{ r}3$

$$\begin{array}{r} 4\text{ r}3 \\ 5\overline{)23} \\ -20 \\ \hline 3 \end{array}$$

The **remainder** is the amount left over when a number cannot be divided evenly. You can multiply and subtract to find the remainder.

Complete Exercises 1–5 to find the remainder.

1. $\begin{array}{r} 6 \text{ r}\boxed{} \\ 5\overline{)32} \\ -30 \\ \hline \boxed{} \end{array}$

2. $\begin{array}{r} 4 \text{ r}\boxed{} \\ 7\overline{)31} \\ -28 \\ \hline \boxed{} \end{array}$

3. $\begin{array}{r} 9 \text{ r}\boxed{} \\ 4\overline{)39} \\ -\boxed{} \\ \hline \boxed{} \end{array}$

4. $\begin{array}{r} 6 \text{ r}\boxed{} \\ 3\overline{)19} \\ -\boxed{} \\ \hline \boxed{} \end{array}$

5. $\begin{array}{r} 6 \text{ r}\boxed{} \\ 9\overline{)62} \\ -\boxed{} \\ \hline \boxed{} \end{array}$

Find the quotient.

6. $5\overline{)\,3\,\,2\,}$

7. $5\overline{)\,1\,\,9\,}$

8. $3\overline{)\,2\,\,2\,}$

9. $2\overline{)\,1\,\,7\,}$

Division Procedures

What is 47 ÷ 3? Remember these steps:

Divide   **M**ultiply   **S**ubtract   **C**ompare and   **B**ring down

$\begin{array}{r} 1\\ 3\overline{)47}\\ -3\\ \hline 1 \end{array}$	**Divide** the tens. **Multiply.** **Subtract.** **Compare.**	$\begin{array}{r} 1\\ 3\overline{)4} \end{array}$ $3 \times 1 = 3$ $4 - 3 = 1$ $1 < 3$
$\begin{array}{r} 1\\ 3\overline{)47}\\ -3\downarrow\\ \hline 17 \end{array}$	**Bring down?**	Yes. Bring the 7 down to make 17.
$\begin{array}{r} 15\\ 3\overline{)47}\\ -3\\ \hline 17\\ -15\\ \hline 2 \end{array}$	**Divide** the ones. **Multiply.** **Subtract.** **Compare.**	$\begin{array}{r} 5\\ 3\overline{)17} \end{array}$ $3 \times 5 = 15$ $17 - 15 = 2$ $2 < 3$
$\begin{array}{r} 15\ r2\\ 3\overline{)47}\\ -3\\ \hline 17\\ -15\\ \hline 2 \end{array}$	**Bring down?**	Nothing is left to bring down. If the number left over is smaller than the divisor, record it as the remainder.

Solve. Use the D, M, S, C, B steps shown above.

1. $5\overline{)65}$ 2. $3\overline{)52}$ 3. $2\overline{)51}$ 4. $4\overline{)91}$

Placing the First Digit in the Quotient

When you begin a division problem, you must first decide where to place the first digit of the quotient.

Example A

tens ones	Divide the tens.	
2 3)7 1 −6↓ 1 1	**M**ultiply. **S**ubtract. **C**ompare. **B**ring down?	$3 \times 2 = 6$ $7 - 6 = 1$ $1 < 3$ Yes. Bring down 1 to make 11.
2 3 r2 3)7 1 −6 1 1 − 9 2	Divide the ones. **M**ultiply. **S**ubtract. **C**ompare. **B**ring down?	$3 \times 3 = 9$ $11 - 9 = 2$ $2 < 3$ Nothing is left. If there is a remainder, record it.

Example B

tens ones	
4)3 9	$3 < 4$, so ask yourself how many 4's are in 39.
9 r3 4)3 9 −3 6 3	Divide the ones. **M**ultiply. **S**ubtract. **C**ompare. **B**ring down? Nothing is left. If there is a remainder, record it.

Write an **x** where the first digit in the quotient should be placed.

1. 5)3 4
   −3 0
      4

2. 3)7 4
   −6 ↓
    1 4

3. 2)1 9
   −1 8
     1

4. 4)5 3
   −4 ↓
    1 3

5. 5)8 3
   −5 ↓
    3 3

Find the quotient.

6. 3)2 9

7. 2)4 7

8. 7)1 9

9. 5)7 5

Problem-Solving Strategy

Guess and Check

There are 87 marbles. After the marbles are divided evenly into bags, there are 2 marbles left over. There are fewer than 10 bags. How many bags are there? How many marbles are in each bag?

Guess:	3 Bags?	4 Bags?	5 Bags?
Check: Remember, there must be 2 marbles remaining.	$$\begin{array}{r} 29 \\ 3\overline{)87} \\ -6 \\ \hline 27 \\ -27 \\ \hline 0 \end{array}$$ No remainder	$$\begin{array}{r} 21\ r3 \\ 4\overline{)87} \\ -8 \\ \hline 07 \\ -\ 4 \\ \hline 3 \end{array}$$ Remainder of 3	$$\begin{array}{r} 17\ r2 \\ 5\overline{)87} \\ -5 \\ \hline 37 \\ -35 \\ \hline 2 \end{array}$$ Remainder of 2! You found the answer.

There are 5 bags with 17 marbles in each bag, and 2 marbles are left over.

There are 82 cookies. When the cookies are divided evenly on plates, there are 4 cookies left over. There are fewer than 10 plates. How many plates are there? How many cookies are there on each plate?

Hint: Guess numbers between 0 and 9. You may need to guess more than three times.

Guess:	____ Plates?	____ Plates?	____ Plates?
Check: Remember, there must be 4 cookies remaining.			

Dividing Three-Digit Numbers

These examples show how to divide a three-digit number.

Example A

$\begin{array}{r} 1 \\ 3{\overline{)529}} \\ -3{\downarrow} \\ \hline 2 \end{array}$	**D**ivide the hundreds. **M**ultiply. **S**ubtract. **C**ompare. **B**ring down?
$\begin{array}{r} 17 \\ 3{\overline{)529}} \\ -3 \\ \hline 22 \\ -21{\downarrow} \\ \hline 19 \end{array}$	**D**ivide the tens. **M**ultiply. **S**ubtract. **C**ompare. **B**ring down?
$\begin{array}{r} 176\ r1 \\ 3{\overline{)529}} \\ -3 \\ \hline 22 \\ -21 \\ \hline 19 \\ -18 \\ \hline 1 \end{array}$	**D**ivide the ones. **M**ultiply. **S**ubtract. **C**ompare. **B**ring down? If there is a remainder, record it.

Example B

$6{\overline{)159}}$	There are not enough hundreds to divide.
$\begin{array}{r} 2 \\ 6{\overline{)159}} \\ -12{\downarrow} \\ \hline 39 \end{array}$	**D**ivide the tens. **M**ultiply. **S**ubtract. **C**ompare. **B**ring down?
$\begin{array}{r} 26\ r3 \\ 6{\overline{)159}} \\ -12 \\ \hline 39 \\ -36 \\ \hline 3 \end{array}$	**D**ivide the ones. **M**ultiply. **S**ubtract. **C**ompare. **B**ring down? If there is a remainder, record it.

Write an **x** where the first digit in the quotient should be placed.

1. $5{\overline{)}}$ | 3 | 4 | 9
-3 | 0 | ↓
   | 4 | 9

2. $3{\overline{)}}$ | 7 | 4 | 5
-6 | ↓
   | 1 | 4

3. $2{\overline{)}}$ | 1 | 9 | 8
-1 | 8 | ↓
   | 1 | 1

4. $4{\overline{)}}$ | 5 | 3 | 2
-4 | ↓
   | 1 | 3

Find the quotient.

5. $5{\overline{)}}$ | 3 | 4 | 9

6. $3{\overline{)}}$ | 7 | 4 | 5

7. $2{\overline{)}}$ | 1 | 9 | 8

8. $4{\overline{)}}$ | 5 | 3 | 2

Practicing Division

In some division problems, you must be careful to place a zero in the tens place or the ones place.

Example A

```
     1
4)429
 -4↓
  02
```
Divide the hundreds.
Multiply.
Subtract.
Compare.
Bring down?

```
    10
4)429
 -4↓|
  02|
 - 0↓
   29
```
There are not enough tens to divide. 2 < 4 Write 0.
Multiply.
Subtract.
Compare.
Bring down?

```
   107 r1
4)429
 -4↓|
  02|
 - 0↓
   29
  -28
    1
```
Divide the ones.
Multiply.
Subtract.
Compare.
Bring down? If there is a remainder, record it.

Example B

```
     2
3)721
 -6↓
  12
```
Divide the hundreds.
Multiply.
Subtract.
Compare.
Bring down?

```
    24
3)721
 -6↓|
  12|
 -12↓
   01
```
Divide the tens.
Multiply.
Subtract.
Compare.
Bring down?

```
   240 r1
3)721
 -6↓|
  12|
 -12↓
   01
 - 0
    1
```
There are not enough ones to divide. 1 < 3 Write 0.
Multiply.
Subtract.
Compare.
Bring down? If there is a remainder, record it.

Find the quotient.

1. 3) 3 2 9

2. 4) 9 2 3

3. 5) 5 4 1

4. 6) 7 2 4

Division Patterns to Estimate

Division Patterns

You can find how many fives are in 2,000 by following these steps.

- Rewrite each problem as a multiplication fact.

- Circle all the numbers in the basic fact to see how many zeros to write in the quotient.

- Write the quotient. If there is a zero in the basic fact, the quotient has one less zero than the dividend. Otherwise the quotient has the same number of zeros as the dividend. You can estimate a quotient by using numbers that are easy to divide.

$20 \div 5 = 4$ ⑤×④=⑳

$200 \div 5 = 40$ ⑤×④0 =⑳0

$2,000 \div 5 = ?$ ⑤×④00 =⑳00

So, $2,000 \div 5 = 400$.

Find the quotient. Use a basic fact and a pattern of zeros to help you.

1. $45 \div 9 =$ _____

$450 \div 9 =$ _____

$4,500 \div 9 =$ _____

2. $21 \div 3 =$ _____

$210 \div 3 =$ _____

$2,100 \div 3 =$ _____

3. $81 \div 9 =$ _____

$810 \div 9 =$ _____

$8,100 \div 9 =$ _____

4. $36 \div 6 =$ _____

$360 \div 6 =$ _____

$3,600 \div 6 =$ _____

5. $72 \div 8 =$ _____

$720 \div 8 =$ _____

$7,200 \div 8 =$ _____

6. $12 \div 4 =$ _____

$120 \div 4 =$ _____

$1,200 \div 4 =$ _____

Estimate the quotient.

7. $278 \div 4 = n$

278 is close to _____.

_____ $\div 4 =$ _____

8. $142 \div 7 = n$

142 is close to _____.

_____ $\div 7 =$ _____

9. $357 \div 9 = n$

357 is close to _____.

_____ $\div 9 =$ _____

Zeros in Division

In some division problems, you need to place a zero in the tens place or the ones place of the quotient.

Zero in tens place

H	T	O
2	0	8

$4\overline{)832}$
-8
$0\ 3$
-0
$3\ 2$
$-3\ 2$
0

$4\overline{)8}$ $4 \times 2 = 8$
Write 2
in quotient.

$4\overline{)3}$ $4 \times 0 = 0$
Write 0
in quotient.

$4\overline{)32}$ $4 \times 8 = 32$
Write 8
in quotient.

Zero in ones place

H	T	O	
2	5	0	r2

$3\overline{)752}$
-6
$1\ 5$
$-1\ 5$
$0\ 2$
-0
2

$3\overline{)7}$ $3 \times 2 = 6$
Write 2
in quotient.

$3\overline{)15}$ $3 \times 5 = 15$
Write 5
in quotient.

$3\overline{)2}$ $3 \times 0 = 0$
Write 0
in quotient.

Find the quotient.

1. $3\overline{)614}$

2. $4\overline{)428}$

3. $2\overline{)819}$

4. $7\overline{)913}$

5. $3\overline{)840}$

6. $5\overline{)652}$

Name _____

Practicing Division

You can use play money to practice division with money amounts.

Divide. $5.26 ÷ 2

Step 1	**Step 2**
Divide the $1 bills into 2 equal groups.	Regroup the one $1 bill left over into 10 dimes.

Step 1

Divide the $1 bills into 2 equal groups.

$$\begin{array}{r} \$2. \\ 2\overline{)\$5.26} \\ -\ 4 \\ \hline 1 \end{array}$$

Step 2

Regroup the one $1 bill left over into 10 dimes.

$$\begin{array}{r} \$2. \\ 2\overline{)\$5.26} \\ -\ 4 \\ \hline 12 \end{array}$$

Step 3

Divide the dimes into 2 equal groups.

$$\begin{array}{r} \$2.6 \\ 2\overline{)\$5.26} \\ -\ 4 \\ \hline 12 \\ -12 \end{array}$$

Step 4

Divide the pennies into 2 equal groups.

$$\begin{array}{r} \$2.63 \\ 2\overline{)\$5.26} \\ -\ 4 \\ \hline 12 \\ -12 \\ \hline 06 \\ -\ 6 \\ \hline 0 \end{array}$$

Find the quotient. Use play money to help you. Line up the dollar sign and the decimal point.

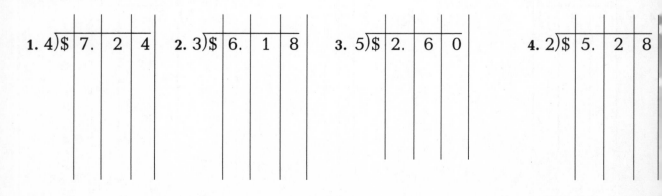

1. $4\overline{)\$\,7.\,2\,4}$ **2.** $3\overline{)\$\,6.\,1\,8}$ **3.** $5\overline{)\$\,2.\,6\,0}$ **4.** $2\overline{)\$\,5.\,2\,8}$

Meaning of the Remainder

The way you interpret the remainder in a division problem depends on the situation.

Jeff took 23 photographs during his trip. He can fit 4 photographs on each page of his album. How many pages will it take to fit all 23 photographs?

Since 5 pages are not enough to fit all the photographs, Jeff needs to use 6 pages.

Julia is making bouquets of flowers. She has 20 flowers in all. If she puts 6 flowers in each bouquet, how many bouquets can she make?

Julia can make 3 bouquets. The remainder is not enough to make another bouquet.

Solve. Tell how you interpret the meaning of the remainder.

1. Sam is packaging cookies for a bake sale. He has 26 cookies. He puts 3 cookies in each bag. How many bags of cookies does he have?

   _____

   _____

2. A group of 42 students are going on a field trip in vans that hold 9 students each. How many vans are needed to transport all the students?

   _____

   _____

3. Becky made 32 clay beads. How many necklaces can she make if she uses 5 beads for each necklace?

   _____

   _____

4. Chris wants to buy cupcakes for 14 people. The cupcakes come 6 to a package. How many packages of cupcakes does Chris need to buy so that each person can have 2 cupcakes?

   _____

   _____

Problem Solving

Account for All Possibilities

In some situations, you may want to divide a number into groups
so that there are exactly the same number in each group.

Mary wants to give the same number of
marbles to each of her 3 friends. She has a
bag of 23 marbles. How many marbles can
Mary give to each friend?

She can give each friend 7 marbles. There
are 2 marbles left over.

In other situations, you may use the remainder to make one
or more of the groups larger.

Megan is making 26 cards for her classmates.
She has 4 days to make the cards. How many
cards should Megan make each day?

She can make 6 cards on 2 of the days,
and 7 cards on 2 of the days.

Solve. Account for all the possibilities.

1. Chen wants to divide his 136
   shells as evenly as possible into
   3 boxes. How many shells can
   he put in each box?

   _____

   _____

2. Joe, Jack, and Henry want to
   divide $4.35 so that they each
   get the same amount of money.
   How much money will each boy
   get? Will there be any money
   left over?

   _____

3. Nan has borrowed a 120-page
   book that is due at the library
   in 7 days. If she wants to finish
   the book in 7 days, how many
   pages should she read each day?

   _____

   _____

4. There are 4 students who are
   working together to write 50
   math problems for their class
   to solve. How many problems
   should each student write?

   _____

   _____

Name _____

Finding the Average

An **average** is one number that best represents all the numbers in a set.

To find the average:

Step 1 Add the numbers.

Step 2 Count the number of addends.

Step 3 Divide the sum by the number of addends.

Examples:

Find the average of 3, 4, 6, and 7.

Step 1 $3 + 4 + 6 + 7 = 20$

Step 2 There are 4 addends.

Step 3 $20 \div 4 = 5$

The average is 5.

Find the average of 5, 7, 8, 9, and 11.

Step 1 $5 + 7 + 8 + 9 + 11 = 40$

Step 2 There are 5 addends.

Step 3 $40 \div 5 = 8$

The average is 8.

Find the average of each set of numbers.

1. 5, 6, and 10

Step 1 $5 + 6 + 10 =$ ____

Step 2 There are ____ addends.

Step 3 ____ $\div 3 =$ ____

2. 2, 4, 5, 6, and 8

Step 1 $2 + 4 + 5 + 6 + 8 =$ ____

Step 2 There are ____ addends.

Step 3 ____ $\div 5 =$ ____

3. 4, 6, 8, and 10

Step 1 ___+___+___+___ = ___

Step 2 There are ____ addends.

Step 3 ___ $\div$ ___ = ___

4. 6, 8, and 10

Step 1 ___+___+___ = ___

Step 2 There are ____ addends.

Step 3 ___ $\div$ ___ = ___

Find the average.

5. 4, 5, 5, 8, 8

____

6. 1, 2, 4, 5

____

7. 4, 7, 10

____

Name _____

Choosing the Operation

Add	Joining groups of different sizes	There are 24 students in Mr. Clark's class and 26 students in Ms. Young's class. How many students are there in both classes? $24 + 26 = 50$ 50 students
Subtract	Taking away or comparing groups	How many more students are in Ms. Young's class than in Mr. Clark's class? $26 - 24 = 2$ 2 more students
Multiply	Joining equal-sized groups	Mr. Clark gave 3 pencils to each of his 24 students. How many pencils in all did Mr. Clark give his students? $24 \times 3 = 72$ 72 pencils
Divide	Finding how many in each group or how many groups	Mr. Clark divided his class of 24 students into 4 equal groups to work on a science project. How many students were in each group? $24 \div 4 = 6$ 6 students

Name the operation you use. Write add, subtract, multiply, or divide. Then solve.

1. Maria used 3 rolls of film on her vacation. Each roll had 36 exposures. How many photographs did she take?

2. Maria bought a photograph album for $12.95. What was her change from a $20 bill?

3. Maria put 6 photographs on each page of her album. How many pages did she use for her 84 photographs of the Rocky Mountains?

4. During a 4-day hiking trip, Maria hiked 5 km, 7 km, 9 km, and 10 km. What was the total length of her hike?

Fractions: Part of a Whole

A **fraction** is a number that names a part of a whole.

	$\dfrac{\text{shaded parts}}{\text{total parts}} = \dfrac{2}{3}$	**Read:** two thirds two out of three two divided by three
	$\dfrac{\text{shaded parts}}{\text{total parts}} = \dfrac{4}{6}$	**Read:** four sixths four out of six four divided by six

Write a fraction for each model.

1. $\dfrac{\text{shaded parts}}{\text{total parts}} = \dfrac{\square}{\square}$

2. $\dfrac{\text{shaded parts}}{\text{total parts}} = \dfrac{\square}{\square}$

3. $\dfrac{\text{shaded parts}}{\text{total parts}} = \dfrac{\square}{\square}$

4. 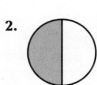 $\dfrac{\text{shaded parts}}{\text{total parts}} = \dfrac{\square}{\square}$

5. $\dfrac{\text{shaded parts}}{\text{total parts}} = \dfrac{\square}{\square}$

6. $\dfrac{\text{shaded parts}}{\text{total parts}} = \dfrac{\square}{\square}$

Write three ways you could read each fraction.

7. $\dfrac{3}{4}$ _____

8. $\dfrac{1}{2}$ _____

9. $\dfrac{2}{3}$ _____

10. $\dfrac{1}{5}$ _____

11. $\dfrac{2}{8}$ _____

12. $\dfrac{8}{9}$ _____

Fractions: Part of a Group

A **fraction** can also name a part of a group.

	$\dfrac{\text{parts shaded}}{\text{number of equal parts}} \begin{array}{l}= 1\\= 2\end{array}$	**Read:** one half one out of two one divided by two
	$\dfrac{\text{parts with filled glasses}}{\text{number of equal parts}} \begin{array}{l}= 4\\= 5\end{array}$	**Read:** four fifths four out of five four divided by five

Write a fraction for each model.

1. $\dfrac{\text{parts with cats}}{\text{number of equal parts}} = \dfrac{\square}{\square}$

2. $\dfrac{\text{parts with stars}}{\text{number of equal parts}} = \dfrac{\square}{\square}$

3. $\dfrac{\text{part with baseballs}}{\text{number of equal parts}} = \dfrac{\square}{\square}$

4. $\dfrac{\text{parts with open umbrellas}}{\text{number of equal parts}} = \dfrac{\square}{\square}$

Write three ways you could read each fraction.

5. $\dfrac{1}{10}$ _____

6. $\dfrac{3}{5}$ _____

7. $\dfrac{2}{3}$ _____

8. $\dfrac{2}{9}$ _____

9. $\dfrac{6}{7}$ _____

10. $\dfrac{5}{10}$ _____

Equivalent Fractions

	1/4		
1/8		1/8	
1/16	1/16	1/16	1/16

1

Fractions that name the same amount are called **equivalent fractions.** $\frac{1}{4}$, $\frac{2}{8}$, and $\frac{4}{16}$ are different names for the same number. So, $\frac{1}{4} = \frac{2}{8} = \frac{4}{16}$, which makes them equivalent fractions.

Complete to find the equivalent fraction.

1. $\frac{2}{3} = \frac{\square}{6}$

2. $\frac{1}{2} = \frac{\square}{8}$

3. $\frac{3}{5} = \frac{\square}{\square}$

4. $\frac{3}{4} = \frac{\square}{\square}$

Color the correct number of parts to show the equivalent fraction. Then write the equivalent fraction.

5.

$\frac{6}{9} = \frac{\square}{\square}$

6.

$\frac{2}{4} = \frac{\square}{\square}$

7.

$\frac{1}{3} = \frac{\square}{\square}$

8.

$\frac{1}{4} = \frac{\square}{\square}$

Comparing and Ordering Fractions

You can compare fractions that have the same denominators.

Compare $\frac{2}{5}$ and $\frac{3}{5}$.
Compare the shaded areas in the fraction models.
$2 < 3$, so $\frac{2}{5} < \frac{3}{5}$.

$\frac{2}{5}$

$\frac{3}{5}$

You can also compare fractions that have different denominators.

Compare $\frac{2}{3}$ and $\frac{1}{2}$.
Compare the shaded areas in the fraction models.
Since $\frac{2}{3}$ has a larger shaded area, $\frac{2}{3} > \frac{1}{2}$.

$\frac{2}{3}$

$\frac{1}{2}$

Compare the fraction models. Write $<$, $>$, or $=$.

1.

$\frac{4}{6} \bigcirc \frac{5}{6}$

2.

$\frac{3}{4} \bigcirc \frac{6}{8}$

3.

$\frac{2}{3} \bigcirc \frac{3}{5}$

4.

$\frac{2}{3} \bigcirc \frac{2}{4}$

5.

$\frac{5}{8} \bigcirc \frac{6}{8}$

6.

$\frac{2}{4} \bigcirc \frac{3}{6}$

7.

$\frac{3}{9} \bigcirc \frac{1}{3}$

8.

$\frac{3}{4} \bigcirc \frac{3}{5}$

9.

$\frac{4}{10} \bigcirc \frac{5}{8}$

Problem-Solving Strategy

Make a Model

Beth made a snack mix for the class party. She used $\frac{1}{2}$ lb of peanuts, $\frac{2}{3}$ lb of pretzels, and $\frac{1}{4}$ lb of dried fruit. List the ingredients in order from greatest to least amount.

You can make a model to help solve the problem. Use fraction bars to model the fractions. Line up the fraction bars so you can tell which bar is the largest and which bar is the smallest.

$\frac{1}{2}$ lb of peanuts

$\frac{2}{3}$ lb of pretzels

$\frac{1}{4}$ lb of dried fruit

The bar for pretzels is the largest and the bar for dried fruit is the smallest. So, the ingredients in order from greatest to least is pretzels, peanuts, and dried fruit.

Make a model to solve.

1. The amusement park has a colorful sidewalk. Starting with the first square, every third square is orange, and every fifth square is green. Draw a model of the first 12 squares of the sidewalk in the space below. What fraction of the first 12 squares are either orange or green?

2. Patti made punch from $\frac{1}{2}$ gallon of grape juice, $\frac{2}{3}$ gallon of kiwi juice, and $\frac{5}{8}$ gallon of papaya juice. Draw a model for each in the space below. Then list the ingredients in order from greatest to least.

_____   _____

_____   _____

Mixed Numbers

A **mixed number** is made up of a whole number and a fraction.

These fraction bars represent a mixed number.

$$\frac{1}{3} \quad \frac{2}{3} \quad \frac{3}{3} \qquad \frac{4}{3} \quad \frac{5}{3} \quad \frac{6}{3} \qquad \frac{7}{3} \quad \frac{8}{3} \quad \frac{9}{3} \qquad \frac{10}{3} \quad \frac{11}{3}$$

There are 3 whole figures shaded.

The last figure is $\frac{2}{3}$ shaded.

So, $3\frac{2}{3}$ figures are shaded.

Or, you can say $\frac{11}{3}$ of the figures are shaded.

Write a fraction and a mixed number for each picture.

1.

_____

2.

_____

3.

_____

4.

_____

Rename each fraction as a mixed number. You may wish to draw a picture.

5. $\frac{3}{2}$ _____

6. $\frac{13}{5}$ _____

7. $\frac{22}{4}$ _____

8. $\frac{11}{3}$ _____

9. $\frac{26}{6}$ _____

10. $\frac{18}{8}$ _____

Adding Like Fractions

When you add like fractions, add only the numerators.

 +   **=**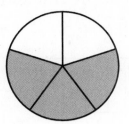

$\dfrac{\text{2 parts shaded}}{\text{5 parts}}$ **+** $\dfrac{\text{1 part shaded}}{\text{5 parts}}$ **=** $\dfrac{\text{3 parts shaded}}{\text{5 parts}}$

$\dfrac{2}{5} + \dfrac{1}{5}$ → Add the numerators. → $\dfrac{2+1}{5}$ → $= \dfrac{3}{5}$
→ Write the denominator. →

Find the sum in Exercises 2–14. Show how you added the numerators.

1. $\dfrac{2}{6} + \dfrac{3}{6} = \dfrac{2+3}{6} = \dfrac{5}{6}$

2. $\dfrac{3}{7} + \dfrac{3}{7} =$ _____

3. $\dfrac{1}{8} + \dfrac{4}{8} =$ _____

4. $\dfrac{2}{5} + \dfrac{2}{5} =$ _____

5. $\dfrac{2}{4} + \dfrac{1}{4} =$ _____

6. $\dfrac{1}{3} + \dfrac{1}{3} =$ _____

7. $\dfrac{4}{10} + \dfrac{5}{10} =$ _____

8. $\dfrac{4}{9} + \dfrac{1}{9} =$ _____

9. $\dfrac{4}{12} + \dfrac{6}{12} =$ _____

10. $\dfrac{2}{8} + \dfrac{1}{8} =$ _____

11. $\dfrac{3}{5} + \dfrac{2}{5} =$ _____

12. $\dfrac{5}{8} + \dfrac{6}{8} =$ _____

13. $\dfrac{3}{9} + \dfrac{3}{9} =$ _____

14. $\dfrac{1}{7} + \dfrac{5}{7} =$ _____

More About Adding Like Fractions

You can add like fractions by making a model and shading parts of it to represent the fractions in the problem.

Find the sum. $\frac{1}{8} + \frac{2}{8} = \underline{\ ?\ }$

Step 1

Since both fractions have an 8 in the denominator, divide a rectangle into 8 equal parts.

Step 2

For the first fraction, $\frac{1}{8}$, shade 1 part out of 8 parts.

Step 3

For the second fraction, $\frac{2}{8}$, shade 2 more parts out of 8 parts.

Step 4

Count the number of shaded parts. There are 3 parts out of 8 parts shaded. So, $\frac{1}{8} + \frac{2}{8} = \frac{3}{8}$.

Create a drawing for each problem and find the sum.

1. $\frac{2}{6} + \frac{3}{6} =$ _____

2. $\frac{2}{8} + \frac{4}{8} =$ _____

3. $\frac{2}{5} + \frac{3}{5} =$ _____

4. $\frac{1}{4} + \frac{2}{4} =$ _____

5. $\frac{1}{3} + \frac{1}{3} =$ _____

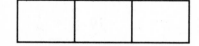

6. $\frac{1}{2} + \frac{1}{2} =$ _____

7. $\frac{3}{10} + \frac{5}{10} =$ _____

8. $\frac{3}{12} + \frac{6}{12} =$ _____

9. $\frac{5}{9} + \frac{2}{9} =$ _____

Subtracting Like Fractions

When you subtract like fractions, subtract only
the numerators.

$$\frac{6}{8} - \frac{2}{8} = \underline{\ ?\ }$$

  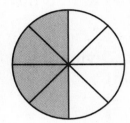

$$\frac{6 \text{ parts shaded}}{8 \text{ parts}} \quad - \quad \frac{2 \text{ parts shaded}}{8 \text{ parts}} \quad = \quad \frac{4 \text{ parts shaded}}{8 \text{ parts}}$$

$\frac{6}{8} - \frac{2}{8}$ → Subtract the numerators. → $\frac{6-2}{8}$ → $= \frac{4}{8}$
$\phantom{\frac{6}{8} - \frac{2}{8}}$ → Write the denominator. →

Find the difference. Show how you subtracted the numerators.

1. $\frac{5}{6} - \frac{3}{6} = \frac{5-3}{6} = \frac{2}{6}$

2. $\frac{6}{7} - \frac{3}{7} =$ _____

3. $\frac{7}{8} - \frac{4}{8} =$ _____

4. $\frac{4}{5} - \frac{2}{5} =$ _____

5. $\frac{2}{4} - \frac{1}{4} =$ _____

6. $\frac{2}{3} - \frac{1}{3} =$ _____

7. $\frac{7}{10} - \frac{5}{10} =$ _____

8. $\frac{4}{9} - \frac{1}{9} =$ _____

9. $\frac{10}{12} - \frac{6}{12} =$ _____

10. $\frac{2}{8} - \frac{1}{8} =$ _____

11. $\frac{9}{10} - \frac{3}{10} =$ _____

12. $\frac{9}{12} - \frac{1}{12} =$ _____

13. $\frac{7}{9} - \frac{5}{9} =$ _____

14. $\frac{5}{5} - \frac{4}{5} =$ _____

15. $\frac{3}{4} - \frac{1}{4} =$ _____

16. $\frac{7}{8} - \frac{2}{8} =$ _____

Name _____

Problem-Solving Strategy

Make a Model

Hector and Melissa each took $\frac{1}{8}$ of the juice in the container. What fraction of the juice was left?

You can *make a model* to find how much of the juice was left.

Step 1

Use a piece of paper. Divide the paper into 8 equal parts.

Step 2

Cut off two pieces, $\frac{1}{8}$ for Hector and $\frac{1}{8}$ for Melissa. There is $\frac{6}{8}$ of the paper left.

So, $\frac{6}{8}$ of the juice in the container was left.

Make a model and solve.

1. John and Peter each took $\frac{2}{5}$ of the ribbon. How much ribbon was left?

2. Ravi is putting up a fence 24 meters long. The fenceposts are 3 meters apart. How many fenceposts does he need?

3. Twelve friends got together to play soccer. In the group $\frac{3}{12}$ wore red shirts and $\frac{4}{12}$ wore blue shirts. The rest wore white shirts. What fraction of the friends wore white shirts?

4. Jim folded a strip of paper in half, then in half again, and then in half once more. When he opened up the folded strip of paper, how many sections did he see?

Adding Mixed Numbers

When adding mixed numbers, first add the fractions and then add the whole numbers.

Find the sum.   **Step 1**   **Step 2**

                       Add the fractions.   Add the whole numbers.

$7\frac{2}{6}$ $7\frac{2}{6}$ $7\frac{2}{6}$
$+4\frac{3}{6}$ $+4\frac{3}{6}$ $+4\frac{3}{6}$
                       $\overline{\frac{5}{6}}$ $\overline{11\frac{5}{6}}$

So, the sum is $11\frac{5}{6}$.

If the problem is written horizontally, you can rewrite the problem vertically and then follow the above steps.

                              $3\frac{1}{7}$ $3\frac{1}{7}$
Write $3\frac{1}{7} + 4\frac{5}{7} = \underline{\ ?\ } \rightarrow$ $+4\frac{5}{7}$ $+4\frac{5}{7}$
                              $\overline{}$ $\overline{7\frac{6}{7}}$

Find the sum.

1. $5\frac{2}{9}$
   $+2\frac{4}{9}$
   $\overline{}$

2. $3\frac{1}{6}$
   $+7\frac{3}{6}$
   $\overline{}$

3. $5\frac{3}{10}$
   $+2\frac{6}{10}$
   $\overline{}$

4. $7\frac{1}{8}$
   $+1\frac{3}{8}$
   $\overline{}$

5. $8\frac{1}{5}$
   $+3\frac{1}{5}$
   $\overline{}$

6. $9\frac{3}{7}$
   $+7\frac{2}{7}$
   $\overline{}$

7. $4\frac{5}{9}$
   $+3\frac{1}{9}$
   $\overline{}$

8. $6\frac{2}{4}$
   $+3\frac{1}{4}$
   $\overline{}$

9. $5\frac{2}{8} + 6\frac{3}{8} = \underline{\ ?\ }$

10. $8\frac{3}{12} + 2\frac{8}{12} = \underline{\ ?\ }$

11. $7\frac{1}{4} + 3\frac{2}{4} = \underline{\ ?\ }$

Subtracting Mixed Numbers

When subtracting mixed numbers, first subtract the
fractions and then subtract the whole numbers.

Find the difference.   **Step 1**   **Step 2**

Subtract the fractions.   Subtract the whole numbers.

$$
\begin{array}{r}
8\frac{5}{6} \\
-\,1\frac{3}{6} \\
\hline
\end{array}
\qquad
\begin{array}{r}
8\frac{5}{6} \\
-\,1\frac{3}{6} \\
\hline
\frac{2}{6}
\end{array}
\qquad
\begin{array}{r}
8\frac{5}{6} \\
-\,1\frac{3}{6} \\
\hline
7\frac{2}{6}
\end{array}
$$

So, the difference is $7\frac{2}{6}$.

If the problem is written horizontally, you can rewrite the
problem vertically and then follow the above steps.

Find the difference.

1. $\begin{array}{r} 5\frac{7}{9} \\ -\,2\frac{4}{9} \\ \hline \end{array}$
　　　2. $\begin{array}{r} 8\frac{3}{6} \\ -\,7\frac{1}{6} \\ \hline \end{array}$
　　　3. $\begin{array}{r} 5\frac{6}{10} \\ -\,2\frac{3}{10} \\ \hline \end{array}$
　　　4. $\begin{array}{r} 9\frac{7}{8} \\ -\,1\frac{3}{8} \\ \hline \end{array}$

5. $\begin{array}{r} 8\frac{4}{5} \\ -\,3\frac{1}{5} \\ \hline \end{array}$
　　　6. $\begin{array}{r} 9\frac{3}{7} \\ -\,7\frac{2}{7} \\ \hline \end{array}$
　　　7. $\begin{array}{r} 4\frac{5}{9} \\ -\,3\frac{1}{9} \\ \hline \end{array}$
　　　8. $\begin{array}{r} 6\frac{2}{4} \\ -\,3\frac{1}{4} \\ \hline \end{array}$

Write vertically and find the difference.

9. $6\frac{3}{8} - 5\frac{2}{8} = \underline{\ ?\ }$　　　10. $8\frac{10}{12} - 1\frac{4}{12} = \underline{\ ?\ }$　　　11. $7\frac{3}{4} - 2\frac{1}{4} = \underline{\ ?\ }$

Relating Fractions and Decimals

You can write a fraction or a decimal to tell what part is shaded.

Model	Fraction	Decimal			Read
		O	T	H	
	$\dfrac{4 \text{ shaded parts}}{10 \text{ parts}}$	0 .	4		four tenths
		O	T	H	
	$\dfrac{25 \text{ shaded parts}}{100 \text{ parts}}$	0 .	2	5	twenty-five hundredths

Complete the table.

	Model	Fraction	Decimal			Read
			O	T	H	
1.		☐ shaded parts ☐ parts	.			
2.			O . T H			
3.			O . T H			
4.			O . T H			
5.			O . T H			

Tenths and Hundredths

You can write a decimal for the part that is shaded.

3 out of 10 equal parts
that are shaded.

Write: 0.3

41 out of 100 equal
parts that are shaded.

Write: 0.41

Write the decimal for the shaded part of each model.

1.

_____

2.

_____

3.

_____

4.

_____

5.

_____

6.

_____

7.

_____

8.

_____

Shade each model to show the decimal amount.

9.

0.7

10.

0.1

11.

0.35

12.

0.99

Equivalent Decimals

Equivalent decimals are different names for the same amount.

0.3 = 0.30

Ones		Tenths	Hundredths
0	.	3	
0	.	3	0

The decimal models show that 0.3 and 0.30 are equivalent decimals. The same amount is shaded in each model.

The place-value chart also shows that 0.3 = 0.30. Both numbers have the digit 3 in the tenths place.

Shade an equivalent amount in each pair of models. Write the decimal for each shaded amount.

1.

2.

_____ = _____   _____ = _____

3.

4.

_____ = _____   _____ = _____

5.

6.

_____ = _____   _____ = _____

Comparing and Ordering

Write these numbers in order from greatest to least:
0.4, 0.35, 0.37, 0.05.

Step 1 Write the numbers in a place-value chart.

Step 2 Write a zero in the empty hundredths place.

Ones		Tenths	Hundredths
0	.	4	**0**
0	.	3	5
0	.	3	7
0	.	0	5

Step 3 Compare the digits in each place, beginning at the left.

Greatest → 0.40, 0.37, 0.35, 0.05 ← Least

Compare. Then write the greater number.

1. 0.2 _____   **2.** 0.25 _____   **3.** 0.36 _____   **4.** 0.5 _____
   0.3   0.27   0.34   0.9

5. 0.37 _____   **6.** 0.02 _____   **7.** 0.38 _____   **8.** 0.04 _____
   0.3   0.10   0.40   0.2

Write <, >, or = in the ◯ .

9.

0.2 ◯ 0.19

10.

0.50 ◯ 0.39

11.

0.25 ◯ 0.2

12.

0.41 ◯ 0.47

13. 0.34 ◯ 0.39   **14.** 0.1 ◯ 0.09   **15.** 0.24 ◯ 0.34   **16.** 0.04 ◯ 0.4

Mixed Decimals

A **mixed number** is made up of a whole number and a fraction.

A **mixed decimal** is a number that is made up of a whole number and a decimal.

$6\frac{1}{10}$

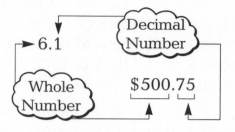

Two whole blocks and three tenths of a block are shaded.

Write: $2\frac{3}{10}$, or 2.3

Write a mixed number and a mixed decimal for the shaded amounts.

1.

_____

2.

_____

3.

_____

4.

_____

Write each mixed number as a mixed decimal.

5. $3\frac{4}{10}$ _____

6. $9\frac{23}{100}$ _____

7. $4\frac{4}{10}$ _____

8. $3\frac{9}{100}$ _____

Write each mixed decimal as a mixed number.

9. 5.6 _____

10. 7.02 _____

11. 3.78 _____

12. 6.09 _____

13. 2.8 _____

14. 6.07 _____

15. 8.02 _____

16. 9.4 _____

Name _____

Problem-Solving Strategy
Make a Table

The students in Ms. Turner's math class participated in a
Metric Olympics. One popular event was the Cotton Ball
Throw. Nick threw the cotton ball 0.91 meter. Noria threw
it 1.45 meters; Caroline threw it 1.39 meters; and Graham
threw it 0.98 meter.

Which student threw the cotton ball
the greatest distance?

The table makes it easy to see who
threw the cottonball the greatest and
the least distances. So, Noria threw
the cotton ball the greatest distance.

COTTON BALL THROW	
Name	**Distance in Meters**
Nick	0.91
Graham	0.98
Caroline	1.39
Noria	1.45

Another popular event was the Paper Plate Discus Throw.
Nick recorded the following distances for this event. Nick–
3.95 meters; Noria–2.80 meters; Caroline–4.15 meters;
Graham–3.81 meters.

1. Make a table for the Paper Plate Discus Throw, arranging
   the distances in order from the least to the greatest.

2. Who threw the paper plate the farthest? _____

3. Who came in second in this event? _____

4. How many students threw the paper plate farther than

   3 meters? _____

Name _____

LESSON
23.1

Modeling Addition and Subtraction

An important thing to remember when adding and subtracting decimals is to keep the place value of numbers lined up.

Find the sum.   0.82 + 0.13   Hint: Let **O** = ones, **T** = tenths,
                                            H = hundredths

0.82

0.13
+

Step 1 Line up the decimal points.

	O	T	H
	0 .	8	2
+	0 .	1	3

Step 2 Add, starting from the right. Work your way to the left.

	O	T	H
	0 .	8	2
+	0 .	1	3
	0 .	9	5

So, 0.82 + 0.13 is 0.95. Subtraction of decimals uses the same two steps.

Use decimal squares to find the sum or difference.

1.

	O	T	H
	0 .	3	
+	0 .	2	
		.	

2.

	O	T	H
	0 .	4	6
+	0 .	3	1
		.	

3.

	O	T	H
	0 .	3	4
+	0 .	6	6
		.	

4.

	O	T	H
	0 .	8	
+	0 .	1	
		.	

5.

	O	T	H
	0 .	9	
−	0 .	2	
		.	

6.

	O	T	H
	0 .	8	7
−	0 .	3	6
		.	

7.

	O	T	H
	0 .	7	
−	0 .	5	
		.	

8.

	O	T	H
	0 .	8	0
−	0 .	7	2
		.	

TAKE ANOTHER LOOK   R139

Adding Decimals

How can you add 1.65 to 0.37?

Remember to keep numbers lined up by place value and their decimal points.

Step 1 Identify the place value of each number.

1.65   0.37

1 is in the ones place.   0 is in the ones place.

6 is in the tenths place.   3 is in the tenths place.

5 is in the hundredths place.   7 is in the hundredths place.

Step 2 Line up each number by its place value.

	O	T	H
	1 .	6	5
+	0 .	3	7
	.		

Step 3 Add, starting from the right. Regroup if you need to.

	O	T	H
	¹1 .	¹6	5
+	0 .	3	7
	2 .	0	2

So, 1.65 + 0.37 equals 2.02.

Rewrite each problem in the place-value grid. Find the sum.

1. $0.8 + 0.5 = n$ 2. $0.9 + 0.6 = n$ 3. $1.67 + 0.58 = n$ 4. $1.46 + 0.17 = n$

5. $0.6 + 1.2 = n$ 6. $1.4 + 1.5 = n$ 7. $1.34 + 0.82 = n$ 8. $1.09 + 0.8 = n$

Subtracting Decimals

How can you subtract 0.58 from 1.72?

Remember to keep numbers lined up by place value and decimal points.

Step 1 Identify the place value of each number.

1.72

1 is in the ones place.
7 is in the tenths place.
2 is in the hundredths place.

0.58

0 is in the ones place.
5 is in the tenths place.
8 is in the hundredths place.

Step 2 Line up each number by its place value.

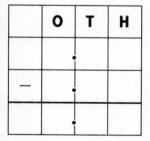

Step 3 Subtract, starting from the right. Regroup if you need to.

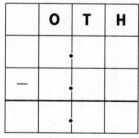

So, $1.72 - 0.58$ equals 1.14.

Rewrite each problem in the place-value grid. Find the difference.

1. $0.7 - 0.1 = n$ **2.** $0.9 - 0.5 = n$ **3.** $1.72 - 0.51 = n$ **4.** $1.91 - 0.85 = n$

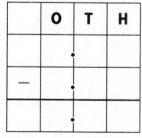

5. $1.7 - 1.5 = n$ **6.** $1.8 - 0.9 = n$ **7.** $1.29 - 0.64 = n$ **8.** $1.13 - 0.88 = n$

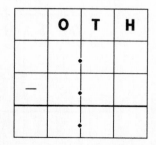

Using Decimals

How can you add 0.3 and 0.45?

Step 1 Write the decimals, keeping the decimal points in a line.

	O	T	H
	0 .	3	
+	0 .	4	5
	.		

0.3

is equivalent to

Step 2 The number 0.3 has fewer digits. Write an equivalent decimal with a zero in the hundredths place so that both numbers have the same number of digits.

	O	T	H
	0 .	3	0
+	0 .	4	5
	0 .	7	5

0.30

Step 3 Add as you would with whole numbers.

Follow the same steps when you subtract.

Rewrite each problem in the place-value grid. Solve.

1. $1.34 + 0.5 = n$ **2.** $1.67 + 0.2 = n$ **3.** $1.2 + 1.32 = n$ **4.** $1.02 + 0.87 = n$

	O	T	H
+			

	O	T	H
+			

	O	T	H
+			

	O	T	H
+			

5. $1.8 - 0.52 = n$ **6.** $2.4 - 1.57 = n$ **7.** $1.20 - 0.56 = n$ **8.** $1.03 - 0.8 = n$

	O	T	H
−			

	O	T	H
−			

	O	T	H
−			

	O	T	H
−			

Problem-Solving Strategy

Write a Number Sentence

Carrie wants to walk to her friend Sue's house, pick her up, and then walk to Joan's house. Then all three girls will walk back to Carrie's house. It is 2.3 miles from Carrie's house to Sue's house. It is 1.6 miles from Sue's house to Joan's house and 3.1 miles from Joan's house to Carrie's house. How many miles will Carrie walk?

Write a number sentence that shows the facts of the problem.

Distance from Carrie's to Sue's house	Distance from Sue's to Joan's house	Distance from Joan's to Carrie's house	Total miles walked	
				2.3
				1.6
				+ 3.1
				7.0
2.3	+ 1.6	+ 3.1	= n	

So, Carrie walked 7.0 miles.

For Problems 1 and 2, use the information above. Write a number sentence to solve.

1. Sue decided to walk home from Carrie's house. What's the total distance she walked today?

2. Joan decided to walk home from Carrie's house. How far did she walk today?

Write a number sentence for each problem. Solve, using problem-solving strategies.

3. A mystery number is added to 8 and then the answer is multiplied by 3. The product is 36. What is the mystery number?

4. Rolanda bought 8 pens for $0.65 each and a notebook for $2.76. How much money did she get back from a ten-dollar bill?

Estimating Sums and Differences by Rounding

You can estimate $2.53 + $1.26 + $0.85 by rounding.

One way to estimate a sum is to round to the ones place and then add.

Step 1 Underline the digits in the ones place.
$$\underline{2}.53 \qquad\qquad \$\underline{1}.26 \qquad\qquad \$\underline{0}.85$$

Step 2 Look at the digit to the right. If the digit is 5 or greater, round to the next higher whole number.
$$\underline{2}.53 \qquad\qquad \$\underline{1}.26 \qquad\qquad \$\underline{0}.85$$
Round 2 up to 3.   Keep as 1.   Round 0 up to 1.

Step 3 Rewrite with zeros in all places to the right.
$3.00 + $1.00 + $1.00 = $5.00

Another way to estimate a sum is to round to the tenths place and then add.

$2.5̲3 $1.2̲6 $0.8̲5
Keep as 5.   Round 2 up to 3.   Round 8 up to 9.
$2.50   + $1.30   + $0.90 = $4.70

Your estimate is either $4.70 or $5.00, depending on how you round.

Estimate each problem twice. Then solve.

Estimate to Ones Place	Estimate to Tenths Place	Exact Answer

1.

```
                                              2 . 3 4
  +  _____        +  _____           + 1 . 6 1
  [_____]         [_____]            [_____]
```

2.

```
                                              3 . 5 4
  +  _____        +  _____           + 1 . 1 7
  [_____]         [_____]            [_____]
```

3.

```
                                              2 . 5 3
  −  _____        −  _____           − 1 . 6 7
  [_____]         [_____]            [_____]
```

Name _____

LESSON
24.1

Linear Measures

When you measure length, you need to first decide which unit to use. Inches, feet, yards, and miles are all linear units which can be used to measure length.

Follow these steps to measure to the nearest inch.

Step 1

Line up one end of the object with the zero mark or the left edge of the ruler.

Step 2

Match the other end of the object with the number to which it is closest.

The pencil is between 3 and 4 inches long. It is closer to 3 inches long. So, to the nearest inch, the pencil is 3 inches long.

Write the measurement to the nearest inch.

1. _____

2. _____

Measure the length to the nearest inch.

3.

_____

4.

_____

5.

_____

6.

_____

7.

_____

TAKE ANOTHER LOOK   R145

Changing Units

To change units, think which measurement is larger. Then think about how the two units relate.

1 foot ▭
1 yard ▭

2 yd = ? ft

Yards are larger than feet. There are 3 feet in every yard. Multiply.
$2 \times 3 = 6$ So, 2 yd = 6 ft.

1 inch □
1 foot ▭

24 in. = ? ft

Inches are smaller than feet. One foot is 12 inches. Divide.

$24 \div 12 = 2$ So, 24 in. = 2 ft.

Larger Unit to Smaller Unit
Number of Miles × 1,760 = ? Yards
Number of Miles × 5,280 = ? Feet
Number of Yards × 3 = ? Feet
Number of Yards × 36 = ? Inches
Number of Feet × 12 = ? Inches

Smaller Unit to Larger Unit
Number of Inches ÷ 12 = ? Feet
Number of Inches ÷ 36 = ? Yards
Number of Feet ÷ 3 = ? Yards
Number of Yards ÷ 1,760 = ? Miles
Number of Feet ÷ 5,280 = ? Miles

1. List the linear units of measurement from least to greatest.

_____

Think about how to change the unit. Write either *larger unit to smaller unit* or *smaller unit to larger unit*. Write *multiply* or *divide* to tell how to change the units.

2. 9 feet to 3 yards   3. 2 miles to 10,560 feet   4. 36 inches to 1 yard

_____   _____   _____

_____   _____   _____

Change the unit. You may use a calculator.

5. 3 yd = _____ ft   6. 72 in. = _____ ft   7. 2 mi = _____ ft   8. 300 ft = _____ yd

Problem-Solving Strategy

Draw a Diagram

Megan is planting rows of tomato plants. The plants have to be spaced 2 feet apart. Her garden is 8 feet by 8 feet. She will start her first row 1 foot in from the edge of the garden. How many rows of tomato plants can she plant?

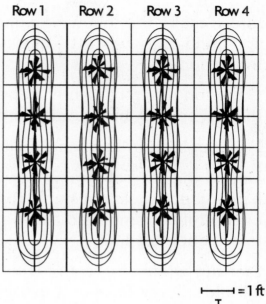

Row 1   Row 2   Row 3   Row 4

= 1 ft

= 1 ft

• To solve, use graph paper to draw a diagram of the garden. Have the length of a square represent 1 foot.

• Since each square represents 1 foot, start the first row 1 space from the edge. Draw the next row 2 feet or 2 spaces over. Continue drawing rows until no more can fit.

So, Megan can plant 4 rows.

Draw a diagram and solve.

1. Megan's tomato plants need to be spaced 2 feet apart from every other tomato plant. She will plant the first plant in each row 1 foot from the edge of the garden. How many tomato plants can Megan fit in her 8 foot by 8 foot garden?

   _____

2. Tom wants to plant a tomato garden like Megan's. He has a square area that is 12 feet by 12 feet. How many tomato plants can he plant?

   _____

3. If Megan and Tom plant marigold plants every 1 foot around the perimeter of their gardens starting at a corner, how many marigold plants will they plant?

   _____

Fractions in Measurements

YOU WILL NEED scissors, paper, ruler

You can make the following paper model to help you identify what the marks on the ruler stand for.

- Cut a long strip of printer paper. Label the left edge 0 and the right edge 1.

- Fold the strip in half lengthwise four times.

- Unfold the strip of paper. You should now have 16 equal sections and 15 folds. Count over 8 folds from 0 and draw a line on that fold. Label this line $\frac{1}{2}$.

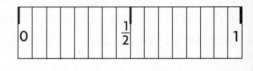

- Starting again at 0, count over 4 folds and 12 folds. Draw lines on the folds. Label the first one $\frac{1}{4}$ and the second one $\frac{3}{4}$.

Use a ruler to measure the length to the nearest $\frac{1}{2}$ inch and to the nearest $\frac{1}{4}$ inch.

1.
2.

3.
4.

5. Draw a chain of 3 large paper clips. Make each $1\frac{3}{4}$ inches long.

6. What is the length of your paper clip chain? _____

Capacity

To change units, think which unit is larger. Then think about how the two units relate.

Smaller Unit ⟶ **Larger Unit**

2 cups = 1 pint	2 pints = 1 quart	4 quarts = 1 gallon

When changing from larger units to smaller units, multiply.

2 gal = _?_ qt

2 × 4 = 8
So, 2 gallons equal 8 quarts.

When changing from smaller units to larger units, divide.

6 c = _?_ pt

6 ÷ 2 = 3
So, 6 cups equal 3 pints.

Find the equivalent measurement by drawing a picture.

1. 12 c = ____ pt	**2.** 4 qt = ____ pt	**3.** 2 gal = ____ c

Weight

Ounce	Pound	Ton
	= 16 oz	= 2,000 lb

Lightest Unit ⟶ Heaviest Unit

To change units, think which unit is heavier. Then think about how the two units relate.

3 lb = ? oz	32 oz = ? lb
Ounces are lighter than pounds. One pound is 16 ounces. Multiply by 16.	Pounds are heavier than ounces. There are 16 ounces in a pound. Divide by 16.
$3 \times 16 = 48$	$32 \div 16 = 2$
So, 3 pounds equal 48 ounces.	So, 32 ounces equal 2 pounds.

Think about how to change the unit. Write either *lighter unit to heavier unit* or *heavier unit to lighter unit*. Write *multiply* or *divide* to tell how to change the units.

1. 2 tons to 4,000 pounds

_____

2. 64 ounces to 4 pounds

_____

3. 6,000 pounds to 3 tons

_____

4. 1 pound to 16 ounces

_____

Circle the more reasonable measurement.

5.

300 T or 300 lb

6.

4 oz or 4 lb

7.

12 oz or 12 lb

Write the equivalent measurement. You may use a calculator.

8. 3 lb = _____ oz   **9.** 7 T = _____ lb   **10.** 50,000 lb = _____ T

Linear Measures

The centimeter, decimeter, and meter are units of length.

| 1 cm | 1 dm | 1 m |

The width of a paper clip is about 1 centimeter (cm).

The width of a computer disk is about 1 decimeter (dm).

The width of a door is about 1 meter (m).

Use the references above to help you decide if the measurement is close to a *cm, dm,* or *m.*

1. the thickness of a calculator

_____

2. the width of a small paperback

book _____

3. the height of a counter _____

4. the length of a large dog _____

5. the thickness of a workbook

_____

6. the height of a big

apple _____

Choose the most reasonable measurement. Write *a, b,* or *c.*

7. width of a softball   **a.** 1 cm   **b.** 1 dm   **c.** 1 m   _____

8. width of a basketball   **a.** 25 cm   **b.** 25 dm   **c.** 25 m   _____

9. height of an average ceiling   **a.** 3 cm   **b.** 3 dm   **c.** 3 m   _____

10. length of a bed   **a.** 200 cm   **b.** 200 dm   **c.** 200 m   _____

Name _____

Decimals and Metric Measures

YOU WILL NEED base-ten blocks, meterstick

The base-ten blocks show how centimeters, decimeters, and meters are related. Each single block is 1 centimeter long. Each tens block is 1 decimeter long. Ten blocks lined up end to end equal 1 meter.

1 centimeter

1 decimeter = 10 centimeters = 1 tens block

$$\frac{1 \text{ cm}}{100 \text{ cm}} = \frac{1}{100} \text{ meter} = 0.01 \text{ meter}$$

$$\frac{1 \text{ dm}}{10 \text{ dm}} = \frac{1}{10} \text{ meter} = 0.1 \text{ meter}$$

Use blocks to help write one measurement as another.

70 cm = __?__ dm = __?__ m

70 cm = 7 dm = 0.7 m

70 cm = 0.7 m

Step 1 Line the centimeter blocks beside the meterstick.

Step 2 Count to find groups of 10 centimeters. There are 7 groups of 10 cm, so there are 7 decimeters.

Step 3 Look at the meterstick. Write the number of centimeters as a numerator over 100.

$\frac{70}{100}$ can be written as 0.70 or 0.7 meter.

Use base-ten blocks and a meterstick to complete the tables.

METRIC LINEAR UNITS			
	Centimeter	Decimeter	Meter
1.	100		
2.	400		
3.	500	50	
4.	60	6	
5.	80		

METRIC LINEAR UNITS			
	Centimeter	Decimeter	Meter
6.	200		
7.	5		0.05
8.	150		
9.	280		
10.	90		

Changing Units

You have to multiply or divide when you change units.

Multiply when you change from a larger unit to a smaller unit.

3 meters = ? dm
3 meters = ? cm

3 m × 10 = 30 dm
3 m × 100 = 300 cm

Divide when you change from a smaller unit to a larger unit.

300 cm = ? dm
300 cm = ? m

300 cm ÷ 10 = 30 dm
300 cm ÷ 100 = 3 m

Circle the larger unit.

1. meter or
   decimeter

2. centimeter or
   decimeter

3. meter or
   centimeter

Would you multiply by 10 or by 100 to change the larger units to
the smaller units? Write × *10* or × *100.*

4. 11 m = ? cm _____   5. 20 dm = ? cm _____   6. 8 m = ? dm _____

Would you divide by 10 or by 100 to change the smaller units to
the larger units? Write ÷ *10* or ÷ *100.*

7. 180 cm = ? dm _____ 8. 200 cm = ? m _____ 9. 800 dm = ? m _____

Fill in the missing measures. You may use a calculator.

	Meters	Decimeters	Centimeters
10.	9		900
11.		100	1,000
12.	17	170	
13.		120	1,200

Name _____

Problem-Solving Strategy

Solve a Simpler Problem

The Problem

Sam grew pumpkins. One of Sam's pumpkin vines was 2.3 meters long. There were 8 pumpkins on the vine. How many centimeters long was the pumpkin vine?

You can solve this problem by breaking it down into parts.

Part 1

Look at the whole-number part first. 2.3 has a whole number of 2. You know that 1 m = 100 cm, so 2 m = 200 cm.

Add the two parts together.
200 cm + 30 cm = 230 cm
Sam's pumpkin vine was 230 centimeters long.

Part 2

Now, look at the decimal part. 2.3 has a decimal part of 0.3. You know that a meter is 100 times larger than a centimeter, so multiply by 100.

$0.3 \times 100 = 30$ cm

Solve a simpler problem.

1. Lisa bought a jump rope that is 180 centimeters long. How many meters long is the jump rope?

2. Ms. Slats bought a new boat. It is 3.4 meters long. How many centimeters long is Ms. Slats's boat?

3. Larry's favorite baseball bat is 95 centimeters long. Moe's favorite baseball bat is 9.8 decimeters long. Whose bat is longer? How many centimeters longer?

4. The school janitor has to set up 320 chairs. He wants 16 or 32 chairs in a row. How many rows can he make?

5. In the past year, Mary's puppy grew 24 centimeters. Loretta's puppy grew 3.2 decimeters. Which puppy grew more in the past year? How many more centimeters?

6. Calvin and Sid both grow tomato plants. Calvin's plant is 11 decimeters tall. Sid's plant is 0.8 meters tall. Who has the taller plant? By how many decimeters?

Capacity

You can measure capacity, or how much liquid a container can hold, by using the units milliliter (mL), metric cup, or liter (L).

 A box the size of a base-ten ones unit holds 1 milliliter (mL).

 250 boxes the size of a ones unit hold a metric cup (250 mL). A small milk carton holds about a metric cup.

 1,000 boxes the size of a ones unit hold a liter (L). A small spring-water bottle also holds 1,000 mL, or 1 L.

Use the references above to help you choose the most reasonable measure. Write a, b, or c.

1. 3 liters = ☐ milliliters

   a. 30 mL   **b.** 300 mL   **c.** 3,000 mL

2. 500 milliliters = ☐ metric cups

   a. 2 metric cups   **b.** 20 metric cups   **c.** 200 metric cups

3. 4,000 milliliters = ☐ liters

   a. 4 L   **b.** 40 L   **c.** 400 L

4. 4 metric cups = ☐ liter

   a. 1 L   **b.** 10 L   **c.** 100 L

5. 3 metric cups = ☐ milliliters

   a. 75 mL   **b.** 750 mL   **c.** 7,500 mL

6. 2 liters = ☐ milliliters

   a. 20 mL   **b.** 200 mL   **c.** 2,000 mL

Write the equivalent measurement.

7. 4 L = _____ · mL

8. 4 metric cups = _____ mL

9. 8 L = _____ mL

10. 8 metric cups = _____ L

11. 500 mL = _____ metric cups

12. 3 L = _____ metric cups

Name _____

Mass

You can measure mass with the units grams (g) and kilograms (kg).

A base-ten ones unit has a mass of 1 gram (g).

If you put 1,000 ones units in a bag, the bag would have a mass of 1 kilogram (kg), or 1,000 grams.

Or you can think of a large container of oatmeal. It has a mass of about 1 kg, or 1,000 g.

Circle each object that has a mass greater than 1 kg.

1. a full-scale airplane   2. a strand of hair   3. a pencil

4. a bed   5. a full backpack   6. a computer disk

Choose the more reasonable measurement. Write *g* or *kg*.

7. a shirt _____   8. a photograph _____

9. a refrigerator _____   10. a big bag of pet food _____

11. a baby brother _____   12. a pair of glasses _____

Circle the more reasonable measurement.

13. spool of thread   a. 10 g   b. 10 kg

14. car tire   a. 20 g   b. 20 kg

15. bag of flour   a. 5 g   b. 5 kg

16. hammer   a. 2 g   b. 2 kg

17. scarf   a. 25 g   b. 25 kg

Time as a Fraction

You can tell time using minutes or fractions.

30 minutes is
one half of
60 minutes.

1:30 can also be read
"half past one."

9:45 can also be read
"a quarter to ten."

4:15 can also be read
"a quarter past four."

15 minutes is
one quarter of
60 minutes.

  or

Draw the hands on the clock to show the time.

1.

3:30

2.

12:45

3.

4:15

4.

half past nine

5.

a quarter to five

6.

a quarter past seven

Write each time in a different way.

7. 10:30

8. 6:45

9. 2:15

_____   _____   _____

10. half past six

11. a quarter to three

12. a quarter past eight

_____   _____   _____

Choosing Customary or Metric Units

You can measure the length of an object using different units.

An inch is a unit of measure in the customary system of measurement.

A centimeter is a unit of measure in the metric system of measurement.

You can say the width of the finger is about $\frac{1}{2}$ inch or about $1\frac{1}{2}$ centimeters.

inches

centimeters

Measure each object in customary and metric units.

1. ⊢_____⊣   2. ⊢_____⊣

_____ in.   _____ in.

_____ cm   _____ cm

3.   Width: Height:

_____ in.   _____ in.

_____ cm   _____ cm

Circle the most reasonable measure for what is shown.

4.

Length of eraser

2 in. or 2 cm

5.

Length of key

5 cm or 5 in.

Problem-Solving Strategy

Write a Number Sentence

You can solve problems by writing a number sentence.

Pierre wants to build a fence around his pool. The fence needs to be 30 feet long and 18 feet wide. What is the perimeter of the fence in yards?

A number sentence can help you solve this problem. To find the perimeter in feet, add all the sides together.

$$30 + 30 + 18 + 18 = 96$$

A second number sentence changes the 96 feet to yards. There are 3 feet in a yard, so divide by 3.

$$96 \div 3 = 32$$

The perimeter of the fence is 32 yards.

Write two number sentences to solve each problem.

1. Patrice needs enough wallpaper border to complete his room. His room is square and measures 15 feet along each side. What is the perimeter of his room in yards?

   _____

   _____

2. Amy needs enough material for a border for a rug. The rug is 9 feet long and 6 feet wide. What is the perimeter of the rug in yards?

   _____

   _____

3. The pool in Kathy's backyard is 5 yards long and 3 yards wide. What is the perimeter in feet?

   _____

   _____

4. Allison bought 4 CDs priced at $12.00 each. The total cost, including tax, was $50.40. How much tax did Allison pay?

   _____

   _____

Temperature

The customary unit for measuring temperature is **degrees Fahrenheit (°F)**.

The metric unit for measuring temperature is **degrees Celsius (°C)**.

Water freezes at 32°F, or 0°C. Water boils at 212°F, or 100°C. Room temperature is about 68°F, or 20°C.

The greater the temperature, the warmer it is.
So, 80°F > 72°F.

Temperature can be compared
by using subtraction. So, 80°F is
8° warmer than 72°F.

$$\begin{array}{r} 80° \\ -\ 72° \\ \hline 8° \end{array}$$

Compare the temperatures. Write < or > to show which is warmer.

1.

50°F ◯ 40°F

2.

10°C ◯ 30°C

3.

80°F ◯ 70°F

Write the temperatures. Write < or > to show which is warmer. Find the difference between the temperatures.

4.

_____ ◯ _____

_____

5.

_____ ◯ _____

_____

6.

_____ ◯ _____

_____

Name _____

Making Equal Areas

You can combine shapes to make different shapes with the
same area. For example,

You can also use shapes to model fractions. The triangle
shows $\frac{1}{3}$ of the trapezoid.

Trace and cut out the pattern-block shapes. Use them to model
each area.

Pattern-Block Shapes

1. Model $\frac{1}{2}$ of ⬡ .

2. Model $\frac{1}{2}$ of ▱ .

3. Model $\frac{1}{3}$ of ⬡ .

4. Model $\frac{1}{3}$ of ⬯ .

5. Model $\frac{1}{6}$ of ⬡ .

6. Model $\frac{2}{3}$ of ⬡ .

Division Patterns to Estimate

In the problem $594 \div 18 = 33$, 594 is the dividend, 18 is the divisor, and 33 is the quotient. To estimate the quotient, follow Steps 1–3.

Step 1	**Step 2**	**Step 3**
Round the divisor to the nearest tens place.	Round the dividend to the nearest multiple of the new divisor.	Divide, using the rounded numbers.
$18 \rightarrow 20$	$595 \rightarrow 600$	$600 \div 20 = 30$

Write the numbers you would use to estimate the quotient.
Then write the estimate.

1. $36 \div 11 = n$

2. $78 \div 13 = n$

3. $56 \div 19 = n$

_____ _____ _____

Write the numbers you would use to estimate the quotient.
Then write the basic fact that helps you find the quotient.
The first one has been done for you.

4. $446 \div 92 = n$

$450 \div 90 = 5$

$45 \div 9 = 5$

5. $52 \div 11 = n$

_____

_____

6. $162 \div 79 = n$

_____

_____

7. $76 \div 24 = n$

8. $272 \div 28 = n$

9. $235 \div 56 = n$

_____ _____ _____

_____ _____ _____

Complete the table.

	Dividend	Divisor	Quotient
10.	40	$\div 20$	_____
11.	_____	$\div 20$	20
12.	4,000	$\div 20$	_____

Dividing by Tens

What is 387 ÷ 60?

$$\begin{array}{r} \text{x} \\ 60\overline{)3} \end{array} \qquad \begin{array}{r} \text{xx} \\ 60\overline{)38} \end{array} \qquad \begin{array}{r} \text{xx}\boxed{?} \\ 60\overline{)387} \end{array}$$

Think: What is the greatest multiple of 60 that can go into 3, 38, or 387?

Write the multiples of 60.

×	1	2	3	4	5	6	7
60	60	120	180	240	300	360	420

- All of the multiples of 60 are greater than 3 and 38.

  So, 60 × 6 = 360 is the greatest multiple of 60 that can go into 387.

- Divide. Place the number 6 in the quotient.

$$\begin{array}{r} \text{xx6 r27} \\ 60\overline{)387} \\ -360 \\ \hline 27 \end{array}$$

---

1. Complete the table to help you solve Exercises 2–5.

×	1	2	3	4	5	6	7	8	9
10	10				50				
20	20								
30		60					210		
40					200	240			
50	50							400	
60				240					540
70			210				490		
80		160						640	
90	90								810

Find the quotient. Use the table to help you.

2. $30\overline{)233}$ 3. $50\overline{)419}$ 4. $20\overline{)189}$ 5. $40\overline{)353}$

Name _____

LESSON
27.3

Modeling Division

You can divide by using models.

What is $35 \div 12$?

Step 1

Count out 35 objects.

Step 2

Create as many groups of 12 as you can.

Step 3

Record your work. Count the number of groups and how many are left over.

$$12\overline{)35} \quad \begin{array}{r} 2 \text{ r1} \\ -24 \\ \hline 11 \end{array}$$

Make a model and find the quotient.

1. $15\overline{)42}$

2. $17\overline{)49}$

3. $25\overline{)103}$

4. $21\overline{)65}$

5. $11\overline{)35}$

6. $31\overline{)128}$

7. $16\overline{)93}$

8. $19\overline{)59}$

9. $24\overline{)84}$

R164   TAKE ANOTHER LOOK

Name _____

Division Procedures

What is 231 ÷ 19? Follow Steps 1–4 when you divide a 3-digit number by a 2-digit number.

Step 1

How many times will 19 go into 2? Zero times. Write an x in the hundreds place.

```
      x
19)231
```

Step 2

How many times will 19 go into 23? One time. Write a 1 in the tens place.

```
     x1
19)231
```

Step 3

Subtract 19 and bring down the 1.

```
     x1
19)231
   -19
     41
```

Step 4

How many times will 19 go into 41? Two times. Write a 2 in the ones place. Subtract.

```
    x12
19)231
   -19
     41
   -38
      3
```

Compare the remainder, 3, with the divisor, 19. The remainder is less than the divisor. Write the remainder as part of the quotient.

So, 231 ÷ 19 = 12 r3.

Find the quotient. Check by multiplying.

1. 11)806

2. 12)185

3. 15)122

4. 16)113

5. 17)414

6. 13)183

7. 18)581

8. 22)119

Problem-Solving Strategy

Write a Number Sentence

Joan and Sam like to play catch with their dogs.
Joan's dog can jump 6 feet into the air to catch a ball.
Sam's dog can jump 68 inches into the air to catch a
ball. How many feet can Sam's dog jump into the air?
Which dog can jump higher into the air?

There are two questions to answer.

- To find how many feet Sam's dog can jump,
  divide the number of inches Sam's dog can
  jump by the number of inches in a foot.

$$\begin{array}{r} 5\ \text{r}8 \\ 12\overline{)68} \\ -60 \\ \hline 8 \end{array}$$

  68 ÷ 12 = 5 r8 So, Sam's dog can jump 5 feet 8 inches.

- To find which dog can jump higher, compare the jumps,
  using the same units.

  Joan's dog can jump 6 feet. Sam's dog can jump 5 feet 8
  inches. So, Joan's dog can jump higher.

Write a number sentence to solve.

1. What if Joan's dog jumps 7 feet
   and Sam's dog jumps 88 inches?
   What is the distance in feet that
   Sam's dog jumps? Which dog
   jumps higher?

   _____

   _____

2. Mark can blow up a balloon
   in 20 seconds. If he blows up
   balloons for 5 minutes, how
   many balloons can he blow up?

   _____

   _____

3. Sid and Charles are brothers.
   Sid is 12 years 2 months old,
   and Charles is 118 months old.
   How many years old is Charles?
   Who is older, Sid or Charles?

   _____

   _____

4. Tom solved the puzzle in
   18 minutes. Shirley solved the
   puzzle in 950 seconds. How
   many minutes did it take
   Shirley to solve the puzzle?
   Who solved the puzzle faster?

   _____

   _____

Correcting Quotients

Divide. $321 \div 38 = n$ $38\overline{)321}$

When you are dividing, sometimes the first digit in the
quotient is too high or too low.

Too High	Too Low	Just Right

$$\begin{array}{r} 9 \\ 38\overline{)321} \\ -342 \\ \hline \end{array}$$

$$\begin{array}{r} 7 \\ 38\overline{)321} \\ -266 \\ \hline 55 \end{array}$$

$$\begin{array}{r} 8 \\ 38\overline{)321} \\ -304 \\ \hline 17 \end{array}$$

Since 342 > 321, the quotient 9 is too high. Use a lesser number.

Since the remainder, 55, is greater than the divisor, 38, the quotient 7 is too low. Use a greater number.

Since the remainder, 17, is less than the divisor, 38, the quotient 8 is just right.

Write *too high, too low,* or *just right* for each estimate. Explain.

1. $\begin{array}{r} 6 \\ 18\overline{)136} \\ -108 \\ \hline 28 \end{array}$ _____

2. $\begin{array}{r} 7 \\ 16\overline{)103} \\ -112 \\ \hline \end{array}$ _____

3. $\begin{array}{r} 8 \\ 17\overline{)139} \\ -136 \\ \hline 3 \end{array}$ _____

4. $\begin{array}{r} 5 \\ 23\overline{)119} \\ -115 \\ \hline 4 \end{array}$ _____

5. $\begin{array}{r} 6 \\ 32\overline{)236} \\ -192 \\ \hline 44 \end{array}$ _____

6. $\begin{array}{r} 7 \\ 28\overline{)170} \\ -196 \\ \hline \end{array}$ _____

Find the quotient.

7. $43\overline{)365}$ 8. $34\overline{)338}$ 9. $52\overline{)379}$ 10. $68\overline{)454}$

Name _____

Making a Circle Graph

A **circle graph** is a graph in the shape of a circle. It shows data as a whole made up of different parts.

A fraction can be used to describe the size of each part.

Write the fraction for the shaded part.

1.

_____

2.

_____

3.

_____

4.

_____

5.

_____

6.

_____

Label each graph.

7. a plate of cupcakes that is $\frac{1}{2}$ chocolate, $\frac{1}{3}$ vanilla, and $\frac{1}{6}$ chocolate chip

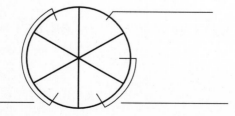

8. a state flag that is $\frac{1}{2}$ green, $\frac{1}{4}$ blue, and $\frac{1}{4}$ white

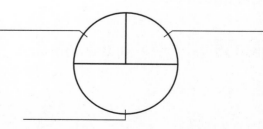

Name _____

Fractions in Circle Graphs

Remember, when you write a fraction for the shaded part, the top number shows the number of parts shaded, and the bottom number shows the number of parts in all.

$\frac{2}{6}$ — number of parts shaded
— number of parts in all

Some other examples:

 $\frac{2}{3}$ $\frac{3}{5}$ $\frac{4}{4}$ $\frac{6}{6}$

Write a fraction for the shaded part. Write a fraction for the unshaded part.

1.   **2.**   **3.**   **4.**

shaded _____   shaded _____   shaded _____   shaded _____

unshaded _____   unshaded _____   unshaded _____   unshaded _____

Write a fraction for each part.

5.   **6.**   **7.**

For Problems 8–10, use the circle graph.

Science Fiction

Comedy

Drama

30 Movie Videos

8. What fraction of the 30 movie videos are comedies? _____

9. Are more of the videos drama or science fiction? _____

10. What fraction represents all 30 videos? _____

Decimals in Circle Graphs

Just as decimals can be used to name parts of a square,
they can be used to name parts of a circle.

    |

| 4 tenths, or 0.4, shaded | 4 tenths, or 0.4, shaded | 10 tenths, or 1.0, shaded | 10 tenths, or 1.0, shaded |

Write a decimal for the shaded part. Then draw a line from the
square to the circle that shows the same decimal.

1. _____

2. _____

3. _____

4. _____

Write the decimal for each part.

5.

6.

7. What decimal tells how many students
chose chocolate milk?

_____

Favorite Milk Flavor of 10 Students

Problem-Solving Strategy

Make a Graph

Two classes went on a field trip to Washington, D.C. Jo and Matt recorded the students' favorite places to visit in a chart, writing both fractions and decimals.

JO'S CLASS (10 STUDENTS)		
	Number of Students	
Place	**Fraction**	**Decimal**
White House	$\frac{1}{10}$	0.1
Washington Monument	$\frac{7}{10}$	0.7
Lincoln Memorial	$\frac{2}{10}$	0.2

MATT'S CLASS (10 STUDENTS)		
	Number of Students	
Place	**Fraction**	**Decimal**
White House	$\frac{3}{10}$	0.3
Washington Monument	$\frac{4}{10}$	0.4
Lincoln Memorial	$\frac{3}{10}$	0.3

To make a circle graph of the data, follow these steps.

Step 1 Write a title for the graph.

Step 2 For each group of data, color the graph to show the fraction or decimal. Use a different color for each group.

Step 3 Label each different-colored section of the graph.

Use the circles below and the information from the tables to make a circle graph for each class.

1. _____

_____

_____

_____

2. _____

_____

_____

_____

Choosing Graphs to Represent Data

To choose the graph that represents a set of data, you must check to see that each part of the data is correctly represented in the graph.

Ms. Tucker's class voted on whether to watch a movie about polar bears, coyotes, or bald eagles. Of the class, $\frac{1}{2}$ voted for bald eagles, $\frac{1}{4}$ for polar bears, and $\frac{1}{4}$ for coyotes. Which graph represents how the class voted?

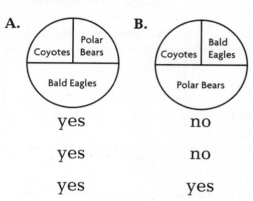

	A.	B.
Does the graph show		
$\frac{1}{2}$ of the votes for bald eagles?	yes	no
$\frac{1}{4}$ of the votes for polar bears?	yes	no
$\frac{1}{4}$ of the votes for coyotes?	yes	yes

So, Graph *A* shows how the class voted.

Read the description of the data, and then answer the questions.

Mrs. Mahajan's class brought in cookies for a bake sale. Of the cookies brought in, $\frac{1}{4}$ were molasses, $\frac{1}{2}$ were chocolate chip, and $\frac{1}{4}$ were peanut butter.

Does the graph show that

1. $\frac{1}{4}$ of the cookies were molasses? _____ _____

2. $\frac{1}{2}$ of the cookies were chocolate chip? _____ _____

3. $\frac{1}{4}$ of the cookies were peanut butter? _____ _____

Which graph represents the cookies brought in for the bake sale? _____